PLANT DISEASE CLINIC
RM 135 FORBES

SYMPTOMS OF VIRUS DISEASES IN PLANTS

Third edition (Revised)

CENTRE FOR AGRICULTURAL
PUBLISHING AND DOCUMENTATION
WAGENINGEN 1978

SYMPTOMS OF VIRUS DISEASES IN PLANTS

WITH INDEXES OF NAMES OF SYMPTOMS IN
ENGLISH, DUTCH, GERMAN, FRENCH,
ITALIAN, AND SPANISH

L. BOS

RESEARCH INSTITUTE FOR PLANT PROTECTION,
WAGENINGEN, THE NETHERLANDS

First published 1963
Second impression 1964 (with corrections)
Second edition 1970 (revised)
Third edition 1978 (revised)

ISBN 90-220-0658-1

Cover design and typography: Pudoc, Wageningen

Printed in The Netherlands

Contents

Preface to the third edition

It was a pleasure to hear so soon from the publisher that this book on such a specialized subject as symptoms of virus diseases needed another edition.

I have carefully gone through the text, that had been completely rewritten for the second edition, and made several minor and major amendments. It has also been updated by adding reference to some 80 publications, which have appeared since 1970.

Although a descriptive field such as symptomatology may not seem to change much, several new and more typical examples of the effects of viruses on plants have become available. Here and there some reorientation was needed because of the increase in number of diseases originally thought to be caused by viruses but now attributed to small filterable micro-organisms. Biochemical research and electron microscopy have increased our knowledge of the nature of symptoms. Therefore, the section on virus particles and inclusion bodies required further elaboration.

Fortunately, this time a number of illustrations in colour could be added. Good pictures often express more than pages of text and several effects of viruses on plants are highly attractive. I regret that the number of figures remains limited for economy.

Critical remarks and suggestions are still welcome.

Wageningen, September 1977 L. Bos

Preface to the second edition

In 1963 the first edition of this book was published for the express purpose of supplying a short general survey of the effects of virus diseases on plants and to draw attention to an interesting field of biological research. It attempted also to eliminate the confusion about the meaning of certain terms describing symptoms.

Although the final interpretation and definition of terms was my own responsibility, they had been agreed upon by many outstanding plant virologists all over the world, who had checked copies of the first manuscript and with many of whom I had personally discussed the subject. Special mention was made of: Dr C. W. Bennett, Salinas, Calif.; Dr R. Bovey, Nyon, Switzerland; Dr M. Chessin, Missoula, Mont.; Dr R. W. Fulton, Madison, Wisc.; Dr P. Grancini, Bergamo, Italy; Dr F. O. Holmes, New York; Dr E. Köhler, Brunswick, West Germany; Dr P. Limasset, Montpellier, France; Dr O. Lovisolo, Turin, Italy; Dr G. Nyland, Davis, Calif.; Dr F. Pelet, Nyon, Switzerland; Dr G. S. Pound, Madison, Wis.; Dr A. F. Ross, Ithaca, N.Y.; Dr H. H. Thornberry, Urbana, Ill.; and Dr H. A. Uschdraweit, Berlin-Dahlem, Germany.

Some eight years have elapsed since the first edition was prepared and a wealth of new information has become available, especially on internal symptoms and about the direct 'physical' participation of the virus particles. Hence the present version has been almost entirely rewritten and considerably extended, among others with a Spanish index of names of symptoms and with a subject index. It contains only a few of the original illustrations. However as stressed in the preface to the first edition, the book is not a comprehensive monograph but only contains selccted examples.

I again acknowledge the encouragement, suggestions and comments received from many colleagues at home and abroad. Special thanks are due to Dr R. Bovey and Dr E. Pelet for revising the French index, to Dr P. Grancini and Dr O. Lovisolo for revising the Italian index, and to Dr K. Schmelzer, Aschersleben, East Germany, for checking the German list. Dr M. Rubio-Huertos, Madrid, Spain, has kindly added a completely new index of Spanish synonyms.

Several virologists have again generously supplied photographs and publishers have given permission to reproduce them. Acknowledgments are given under the figures concerned.

I am greatly obliged to Pudoc (Centre for Agricultural Publishing and Documentation), Wageningen, for the able technical help that has already played a considerable role in the success of the first two impressions, and especially to Dr E. Meijer Drees and Mr J. C. Rigg for their careful editing and help with the English.

Miss M. P. Schor helped in making the indexes and Mr C. A. Koedam and Mr C. F. Scheffel prepared the photographs.

Finally I want to thank my director Dr J. G. ten Houten, Institute of Phytopathological Research, for allowing me the time and assistance necessary to prepare this revised edition.

Wageningen, 1970 L. Bos

Introduction

The importance of viruses in initiating plant diseases is now an accepted fact. The losses they cause every year in crops run into millions. More than ever before, discoveries about virus diseases, both for practical reasons and for pure scientific interest, are topical.

Viruses differ from pathogenic micro-organisms in various respects. Usually micro-organisms can readily be grown on artificial media and studied outside their hosts. Viruses may be defined as submicroscopic and filterable incitants of infectious disorders that can pass through porcelain filters, with pores small enough to retain bacteria, and that are not able to multiply in cell-free artificial media. However, a rapidly increasing number can be isolated from their hosts and studied in vitro. These viruses are now being described in macromolecular terms as nucleoproteins. They sometimes even consist of naked nucleic acid, and sometimes have an extra envelope in addition to the normal protein coat. Their nucleic acid genome is of one type, either RNA or DNA, with a molecular weight of less than 3×10^8 daltons. Hence, genetic information is limited. The viroids discovered in 1972 consist only of short RNA strands of circa 10^5 daltons, hardly sufficient to code for one gene.

Structure and even 'filterability' have not been demonstrated in viruses that so far are not transmittable with sap from plant to plant or from vector insect to insect. Several diseases, formerly accepted to be caused by viruses, have since 1967 proved to be due to filterable mycoplasmas and even to Rickettsia-like or other tiny microbes. These discoveries demonstrate the inadequacy of the usual definition of virus. The virus nature of a pathogen is not well-established unless it has been characterized physico-chemically. Unlike micro-organisms, viruses have no metabolism of their own, so that the relationships with their hosts are entirely different.

For a long time, viruses were known and distinguished only by host reaction, and several can still only be studied in that way. So at first, virus pathology predominated; now the balance has swung to the other extreme, as more and more viruses are isolated, purified and studied in vitro.

Modern virology has concentrated on the chemistry and physics of viruses rather than on their pathology, for they are useful tools in molecular biology. Emphasis is then often on molecular structure, especially of the protein coat and on base composition of the nucleic acid. However, there is increasing awareness that viruses are polycistronic (containing several genes) and that only part of the total genetic information comprised in the viral nucleic acid is expressed in protein coat configuration and thence in such 'intrinsic' properties as particle morphology, size and serological affinity. What basically counts is nucleotide sequence in the nucleic acid and its perceptible effects on host biology. Thus, symptomatology has regained considerable interest.

Definition and significance of symptoms

Disease *symptoms*[1] are the effects of viruses and other pathogens on growth and development of plants and other organisms. They may be more sharply defined as any perceptible change in the body or its functions, indicating a disturbance in the normal course of the physiological processes. Though this definition seems very simple, its implications are not, as will be discussed in the next Chapter.

There is only an indirect relation between the term symptom and the original literal meaning of the Greek word σύμπτωμα, symptoma (= accident, change, anything that has befallen one). It is derived from συμπίπτειν, sympiptein, which means 'to fall together', and thus somewhat resembles 'to coincide'. Though it actually is too weak to express the close association between disease and symptom, the term is in general use.

In spite of the application of such modern diagnostic methods as serology and the use of the electron microscope, knowledge of symptoms is still essential for a quick orientation, a rapid diagnosis of virus diseases in the field, for virus identification with the help of indicator plants, and for other types of work with test plants (Bos, 1967, 1976). Many viruses can still only be studied by their effects on plants and their way of transmission.

[1] The terms printed in italics are listed in the index with Dutch, German, French, Italian and Spanish equivalents. Italics are only used where a term is defined.

Since viruses are capable of changing almost any organ, influencing almost any function, symptoms constitute a fascinating field of biological research and their study greatly assists understanding of how viruses work.

For the agriculturist the nature of symptoms determines the nature and extent of virus damage in crop production. Because of his interest in the economics, reduction in yield and quality must also be estimated.

Nomenclature of symptoms

For a mutual understanding in research and practical work, names of symptoms must be standardized. In describing the symptoms, a scientific kind of language has gradually developed, but a consistent nomenclature of viruses and virus diseases based on virus classification is still lacking. The viruses are still called by 'common' names, usually based on the most characteristic symptom or complex of symptoms, associated with the disease in combination with the name of the typical host or the host first studied, e.g. barley yellow dwarf virus.

Symptoms can be named in two ways, whether they are symptoms of virus diseases, of other infectious diseases or mere physiological or genetic disorders. The symptoms can be indicated by the final result of the infection (such as 'yellow edge' or 'mottle' of a leaf), or by the process altering the appearance of the diseased organ ('edge yellowing' or 'mottling', respectively). Since conditions are rarely (if ever) static in living material, the process form seems preferable. However the names of symptoms should be short lucid code words so that many authors prefer, for instance, 'mottle' and 'wilt'. Moreover, no acceptable short term exists for the process underlying 'mosaic'.

Consequently both ways of naming symptoms are used in the literature of plant viruses, often in combination.

Finally, there are some terms indicating both process and outcome, such as 'malformation'.

To achieve some standardization in symptom terminology the manuscript of the first edition of this book was sent to several outstanding virologists in various countries with a request for suggestions and critical remarks. One outcome of carefully considering their comments was the six-language index at the end of the book. It is hoped that it will contribute to a uniform international use of terms and so simplify study and interpretation of literature on viruses.

Pathogenesis

Except in diseases caused by environmental factors or inherent defects, symptoms are the result of infection. They seldom occur separately, one symptom being characteristic for the whole process; usually various symptoms occur simultaneously, or they may form a series, one leading to another. The whole cycle of symptoms is called disease.

For a better understanding of the nature of the symptoms to be described in the next chapter, some terms must first be explained. These terms reflect present knowledge of the nature of disease and especially of how the plant gets sick. In *pathogenesis* (the development or evolution of a pathological process or disease) such aspects as origin, sequence, interaction and variability of symptoms are involved.

Definition of disease

Symptoms are perceptible changes in the body or its functions indicating disease. The adjective 'perceptible' will be discussed more fully when dealing with the nature of symptoms and their absence. Here first the concept *disease* has to be defined. Most books on plant pathology agree that it is difficult to circumscribe plant disease. Thus there is considerable variation in definitions given.

In Webster's Third New International Dictionary (1961) the term disease is described as 'an impairment of the normal state of living in animal or plant body or of any of its components that interrupts or modifies the performance of the vital functions, being a response to environmental factors, to specific agents, to inherent defects of the organism, or to combinations of these'. In Webster's New Collegiate Dictionary (1960) it is more briefly described as 'a condition in which bodily health is impaired; sickness'. These definitions leave open the question of what is meant by 'normal state' of the body and by 'health'. Anyhow, functions are abnormal or impeded so that the body, or parts or organs suffer (Gr. πάθος, pathos = a suffering, passion; akin to Gr. πένθος penthos = grief, from Gr. πάσχειν, paschein = to suffer). If normal development or function is completely impossible, death may ensue. This definition reduces, so to say, the question to a statistical problem: all that surpasses normal natural variation is *abnormal*. It should be clear, however, that because irritation by the pathogen may be continuous and the host reacts, disease differs from mere injury, for instance by insects with biting mouth parts.

The next question is: when is life threatened or when do the organism or its parts suffer? This is also largely a statistical problem, because for plants the pain concept does not hold and suffering can only be conceived as reduced ability to compete, reduced size or reduced vigour expressed in terms of shorter life, reduced yield, or reduced resistance to adverse conditions. In plant life, 'suffering' (as judged by man) often has a subjective connotation as something harmful, injurious or disadvantageous, and this is usually explained by man as being less productive, as measured in product weight or financial outcome. It should be further emphasized here that the term disease in anthropomorphic. Biologically, disease is a normal phenomenon. Pathogenic organisms have as much right to live as more directly useful ones. They are essential for the maintenance of biological equilibrium on earth.

A good example to illustrate this situation is flower colour breaking in tulip caused by tulip breaking virus (Plate 1). Around 1634–1637 fortunes were paid for the bulb of a nicely 'broken' tulip (Fig. 1). Thus at first the deviation was considered useful. Only later was it discovered that 'unbroken' varieties could get lost by infection. We now know that the colour deviations gradually intensify and that size of plant and flower and yield of bulbs are reduced.

The beautiful leaf variegation of *Abutilon striatum* 'Thompsonii' due to *Abutilon* mosaic virus (Fig. 2) poses an interesting problem. In temperate regions, this plant is widely grown indoors and, although infected, it does not suffer. In fact, the virus has been responsible for its wide distribution as a vegetatively propagated ornamental. Under these conditions the virus does not manifest its infectivity for other plants because it is not transmitted by sap and the insect vector has a limited distribution. Thus the question arises: is *Abutilon* mosaic a disease?

Origin of symptoms

Symptoms are here defined as perceptible changes in the body or its

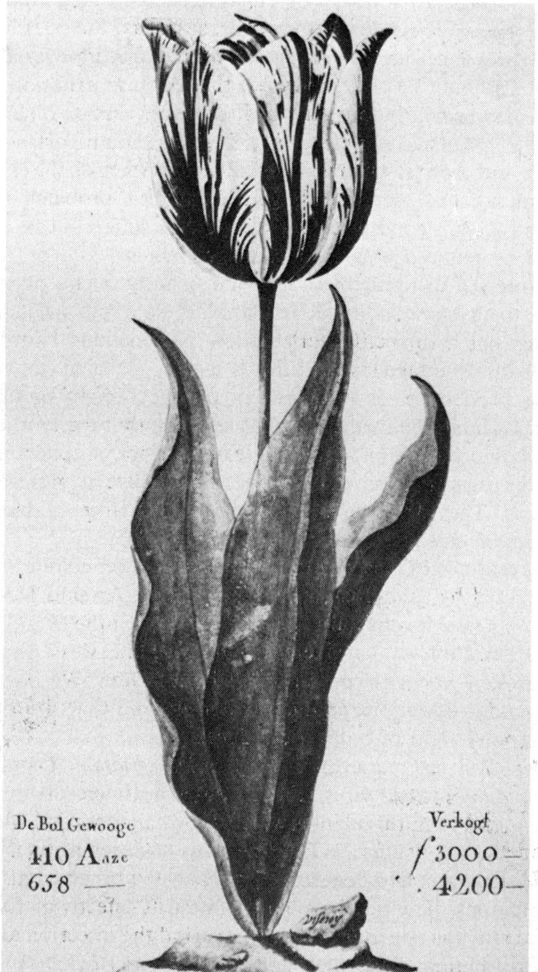

Fig. 1 Tulip with colour breaking, of which one bulb was sold in 1637 for the fabulous price of 4200 guilders and another for 3000 guilders. Many such tulips were depicted by Dutch painters of 17th century. Black-and-white reproduction of a painting by Cos (1637).

Fig. 2 Mosaic in *Abutilon striatum* with *Abutilon* mosaic virus.

functions. This assumes a reaction of the victim, though the mere presence of virus particles, fungal hyphae, spores and fruiting bodies, or bacterial cells in or on the diseased host may be an important characteristic of a certain disease.

Viruses, however, are not visible to the naked eye, and with the light microscope only in the few cases where they aggregate as or participate in 'inclusion bodies'. They 'act' only inside the living host cells and their multiplication is a product of host activity as well. Thus viruses are unique among plant pathogens (pathogen = disease incitant). Let us, therefore, start at the beginning of virus-plant relationships to get a good understanding of the origin of symptoms of virus diseases in plants, and thus of *pathogenesis.*

In contrast to parasitic organisms, viruses can get into the plant only through extremely small wounds not killing the cells to be infected. Such wounds may result, for instance, from breaking of epidermal hairs by contact between diseased and healthy plants, or be caused by probing or feeding insects carrying the virus. The purposeful introduction by man of virus into living cells is called *inoculation.* It can be done either mechanically, as by rubbing infective sap and an abrasive on a leaf, or with insects, mites, nematodes or certain fungi.

Only if the inoculated cells are susceptible to the particular virus (i.e. when the viral genetic code fits the normal biochemical pattern of the host) can the virus become established in the cell's metabolism and take effect. Many details of this process remain to be discovered. Establishment of the virus and its taking effect is called *infection.*

Inside the locally infected plant, the advance of the process is usually continuous. From the entrance site the virus may spread as a systemic infection to other cells and so affect surrounding tissues and usually the entire plant and any newly formed organs.

The term 'infection' is often used to indicate the communication of an infectious disease from one organism to another. It should be kept in mind, however, that virus infection *sensu stricto* occurs inside the host cells and is not accomplished by man or any vector but by the virus itself. Infection *sensu stricto* is a prerequisite to disease, it initiates disease. What occurs outside the host is solely *contamination* by a contagion, bringing the victim into contact with the contagious agent, and it differs from inoculation. Man cannot infect *(sensu stricto)* a plant or other host: this is the exclusive privilege of the pathogen.

If the inoculated cell is not susceptible, contamination or inoculation does not result in infection. Total lack of *susceptibility* is called *immunity*; the plant passively resists attack.

In susceptible plants the virus initiates abnormal processes leading to virus multiplication, and biochemical and physiological changes of the host. Much is still unknown about the way viruses get established in the host's biochemistry, but abundant information is available on the abnormalities resulting from infection (see p. 41, biochemical and metabolic changes). This information has revealed that virus symptoms are mainly an expression of aberrant cell metabolism. These aberrations start in the inoculated cells, extend to surrounding tissue, and may finally affect the entire plant. Initially they can be detected cytologically but may later become visible to the naked eye. The aberrations may form a sequence, ultimately modifying large parts of the plants and, if the host suffers from them, constituting a *disease*.

The group of symptoms or pathological phenomena caused by a virus or other pathogen is known as the *syndrome* (Gr. συν, syn = with, together; δράμειν, dramein = to run). The term clearly indicates the dynamic nature of disease: it is composed of several phenomena developing side by side, more or less parallel, or interdependent. The syndrome is the clinical picture of the disease, though the disease starts much earlier, at the moment of infection. The period between the entry of the virus and first symptoms is the *incubation period* (literally 'brooding' period, from L. incubare = lie upon, hatch; in = in or on, cubare = lie down, lie upon).

Infection, evident from detectable virus multiplication, does not always lead to visible symptoms (p. 30, absence of symptoms). It may be present without showing up. The outcome depends on type or strain of the virus, on external conditions, and on the genotype of the host. For agriculture this is very important, as viruses themselves are hard to control, whereas growing conditions can be considerably influenced and plant genotype can be changed by breeding.

Thus the severity of the pathological effects in a susceptible host depends partly on its *sensitivity*, its capacity to react visibly and its vulnerability. A low sensitivity is equivalent to *tolerance*, which is characterized by mild symptoms; the virus may multiply normally. Extreme tolerance results in imperceptible host reaction, in *inapparency* of the infection. Sensitivity may also be extremely high, so that the originally infected cells or their surroundings are rapidly killed. Such *hypersensitivity* may prevent further spread of the virus inside the plant.

It is often hard to distinguish between sensitivity and susceptibility and to determine their role in a given syndrome. Both are quantitative characters of the host. *Susceptibility* refers to the ease with which infection starts and spreads within the host. Severe symptoms may be due to easy local infection followed by a rapid systemic infection or, at a low rate of infection and virus multiplication, to a high sensitivity of the invaded tissue. So the number of local lesions formed after inoculation may indicate the degree of susceptibility of a certain host (at least of susceptibility to local infection), whereas the size or nature of the lesions indicate the degree of sensitivity. But in sensitivity, the rapidity of virus multiplication in the tissue and the speed of its spread to surrounding cells may also play a role.

This all demonstrates how susceptibility to virus infection is regulated by various factors. Usually the share of such factors, as the ease of establishment, the ease and speed of multiplication, the ease and speed of virus transport inside the plant and plant sensitivity are not known. Therefore in practice, for instance among plant breeders, the term *resistance* is used to cover all host factors that cause a plant to escape infection or to withstand infection without adverse effects. Low sensitivity or low susceptibility may both be involved. As the term resistance indicates lack of knowledge about the cause of the phenomenon, its use should be avoided wherever possible, or it should be specified to prevent misunderstanding. The term is usually used in the forms 'resistance to disease' or disease resistance, which refer only to mildness of symptoms. Sometimes it is applied more specifically as 'resistance to the virus or other pathogen', referring more to restricted or retarded entry, multiplication and internal spread of the pathogen. So a plant reasonably resistant to a certain virus disease is not necessarily resistant to the virus concerned but may be highly tolerant. By breeding and selecting for resistance to visible disease in potatoes and other crops, tolerance to virus infection and, consequently, infection with 'latent' viruses (p. 32) has become widespread. Extreme or total resistance to the pathogen equals *immunity*, as does insusceptibility (p. 17). In practice plants may escape disease and show resistance to natural infection also by hypersensitivity (p. 21).

Similar problems are associated with the term *predisposition*, indicating 'disease proneness' of a host, with emphasis on the environmentally and developmentally conditioned susceptibility (Yarwood,

1959). It should be well understood that reference to host capacity to be diseased is at variance with the term susceptibility as defined earlier, because predisposition involves sensitivity too. Both terms, resistance and predisposition, indicate lack of accurate knowledge of the individual part played by susceptibility to the pathogen and by sensitivity of the host. Sensitivity is often covered by resistance, also when erroneously speaking of 'susceptibility to disease'. The fallacy of doing so is indicated by the impossibility of speaking of 'susceptibility to drought or frost injury or to deficiency' where sensitivity is meant. In defining the term predisposition, environmental conditions are always included but they play a similar important role in determining resistance.

So I propose here to use predisposition and resistance as counterparts in describing host relationships to disease; the first referring to disease proneness, the latter to capacity to resist disease. Both terms should be avoided when susceptibility (as measured by the amount of pathogen) and sensitivity (as measured by the severity of host reaction) can be clearly distinguished. Of course, with viruses the situation is often more difficult to judge than with fungi, where the amount of mycelium, spores or fruiting bodies can be easily estimated and the role of the pathogen can be more easily determined.

With plant viruses in their hosts, susceptibility to infection and sensitivity may act completely independently. This phenomenon is obvious with cucumber green mottle mosaic virus in cucumber, where tolerance may be accompanied by different levels of virus content ranging from high to very low, and low virus content may even lead to severe symptoms depending on the host cultivar (Čech & Branišová, 1976).

Viruses multiply most rapidly in young growing parts of plants because they depend entirely on the physiological activity of the host, presumably especially on its nitrogen metabolism. But, as Esau (1948) has pointed out, though mature parts are less sensitive and the transport of assimilates (and thus that of virus particles) towards them is restricted, they too may develop symptoms after inoculation. Nevertheless, for the same reason that led Bercks (1951) and Beemster (1958) to distinguish *mature plant resistance* to viruses, there is a certain resistance of mature parts to virus infection. It differs considerably according to the kind of virus.

Sequence of symptoms

Diseases do not just occur; they consist of a sequence of various stages in the course of their development, a succession of events or modifications, one usually leading to the other. In other words: the course of a disease is characterized by a sequence of symptoms.

The *local symptoms* occur at or close to the 'locus' or site where the virus has entered. Where a distinct area of diseased tissue becomes visible, this is called a *lesion* (L. laesio = to injure) or more specifically a *local lesion* (Figs 3, 4 and 5 left). It may refer to a discoloration, to a desiccation, or even to the death of the local tissue; examples will be discussed later on. Sometimes local lesions are visible only under the microscope (Helms & McIntyre, 1962), or after staining, e.g. when they are associated with a local accumulation of starch (Fig. 11) (Holmes, 1931; Helms & McIntyre, 1962).

In accordance with general usage the adjective 'local' has been added to stress that the lesions concerned start at the 'locus' of inoculation. It does not refer to the localization of the reaction to a restricted area of tissue, since this is already expressed by the word lesion itself. Lesions may also occur in plant parts systemically invaded by the virus (see below).

Fig. 3 Local lesions in *Solanum demissum* × *S. tuberosum* 'Aquila' (A6) with potato virus A (left) and in *Chenopodium quinoa* with *Passiflora* latent virus (right). (Left, from Dr J. A. de Bokx, Wageningen.)

Local symptoms are most likely to occur after mechanical inoculation with its great number of infection sites. Insects, though causing only a few such sites, may act in the same way, as with bean common mosaic virus in primary leaves of French beans. Holmes (1929) was the first to use necrotic local lesions in *Nicotiana rustica*, and especially in *N. glutinosa*, to measure virus concentrations in an inoculum of tobacco mosaic virus In the latter species the lesions sometimes appear within thirty hours.

Local lesions may gradually enlarge as the virus moves rather slowly from cell to cell, e.g. in the leaf mesophyll (Fig. 4 left). Often such infected tissue rapidly dies, death preventing further virus spreading and leading to 'abortion' of the infection (Fig. 4 right). This *hypersensitivity* can be important as it results in resistance to virus and disease under natural conditions. A good example is the 'field resistance' of bean varieties descending from the North American 'Corbett Refugee' to bean common mosaic virus (Quantz, 1958b). But localization of infection does not necessarily depend on cell necrosis (Shimomura, 1971; Takahashi, 1975). There are other ways for the plant to fight back (p. 27–30).

Local lesions tend to extend into asteroid spots if they reach the veins and then to further spread along the veins (Plate 2). This is because cells along the veins are elongate, so that lesions extend along the veins up to 10 times as rapidly as transversely to them (Köhler, 1964).

Fig. 4 High sensitivity (left) and hypersensitivity (right) in leaves of pea *(Pisum sativum)* after sap inoculation with pea early-browning virus (left) and with the related tobacco rattle virus (right). With pea early-browning virus, lesions continue to enlarge and the virus becomes systemic. (After Bos & van der Want, 1962.)

Usually the virus becomes systemic on reaching the vascular bundles and moving rapidly with the sap stream in the phloem towards the growing parts of the plant. In these previously healthy parts it causes *systemic symptoms*. They will be further discussed below.

As to their chronological order a distinction is sometimes made between *primary* and *secondary symptoms*. Usually the primary symptoms are local lesions. But where these are absent, the first systemic reaction, such as *vein chlorosis* (Fig. 17) has to be considered as such. Sometimes local lesions become visible much later than the first systemic deviations, and the terms local symptom and primary symptom are not equivalent, as a secondary symptom is not always a systemic symptom.

The adjectives 'primary' and 'secondary' have a rather arbitrary meaning. In seed-transmitted diseases they indicate stages of a continuous systemic infection. In primary leafroll of potatoes the 'primary stage' develops in the field soon after infection through aphids as a systemic symptom in the tips of young leaves, whereas the 'secondary stage' (secondary leafroll) develops from the infected tuber early the next season (e.g. Oort, 1959). The terms may also be used to indicate consecutive stages in pathogenesis.

The long-distance transport of virus through the phloem vessels is, together with that of the food, mainly directed towards the growing parts of the plant. A transport of virus through the xylem has only been observed with bean southern mosaic virus.

Once the virus in the phloem vessels has reached the veins of systemically infected leaves, its transport slows down as it has to move from cell to cell into the mesophyll. So it is not surprising that systemic symptoms often start in or close to such veins as vein clearing, vein chlorosis and yellowing (Figs 17 and 18), or vein mosaic (Fig. 19A, B) and vein banding (Figs 5 right and 19E). If the young veins are highly sensitive, *vein necrosis* (Fig. 6) may occur. This is typical of French bean black root characterized by vascular necrosis in root, stem, leaves and pods of certain varieties of French bean under certain conditions, and with some strains of bean common mosaic virus (Grogan & Walker, 1948). In such cases the necrosis may prevent further spread of the virus or make it erratic, as in pea early browning (Bos & van der Want, 1962). The underlying process resembles that of hypersensitivity.

In other cases the virus slowly advances from the veins, especially the bigger ones, inside the mesophyll. It reaches a sufficiently high concentration to become visible only at some distance, soon afterwards decreasing because of host defence mechanisms (p. 27). This causes a *line pattern* (Fig. 19F, G) or an attractive *oak-leaf pattern* (Fig. 21).

Quite often systemic infection starts with the development of limited areas of diseased tissue. Such *systemic lesions* may well be due to a low concentration of virus in phloem contents, which gives rise to a shock-type of reaction restricted to those spots where infectious units succeed in establishing new centres of multiplication (Fig. 20). Physiologically such lesions seem to be identical with local lesions. After some time spots may become necrotic or chlorotic, and may gradually extend some distance along veins or into surrounding tissue, giving rise to attractive patterns (Fig. 6).

Fig. 5 Local (left) and early systemic reaction (right) in leaves of 'White Burley' tobacco with alfalfa mosaic virus.

Most 'persistent' insect-borne viruses, which cannot be transmitted mechanically into parenchymatous tissue, have to be introduced directly by the vector into phloem together with the insect's saliva (e.g. Esau, 1961). Some of them multiply in their vector and are thus introduced into their host in such concentrations that the diseases start with systemic infection. Multiplication of these viruses seems to be restricted to the phloem, hence the resulting abnormalities mainly occur or start in or around the vascular bundles (Fig. 18).

In local symptoms the diseased tissues generally differ in physiology from those in systemic symptoms. When infected they are usually older and further differentiated. Though at the outset local and systemic symptoms sometimes resemble each other, especially in their lesions, they are usually essentially different. This has clearly been demonstrated by Schippers (1963) for bean common mosaic in French

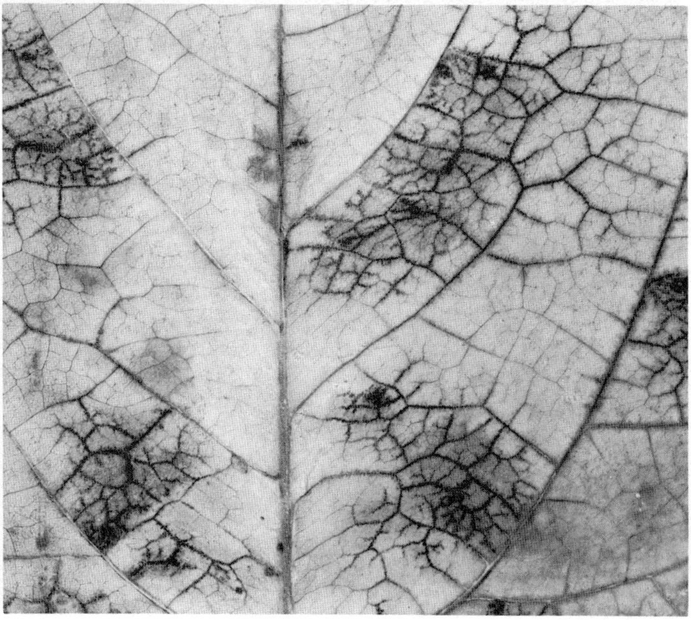

Fig. 6 Systemic vein necrosis in a chlorotic leaf of French bean *(Phaseolus vulgaris)* 'Bataaf' with an Italian strain of *Wisteria* vein mosaic virus.

beans where, according to leaf size and thus to stage of development at the presumed moment of systemic invasion, the leaves show either a coarse mosaic, a fine mosaic, transitional symptoms (brittle leaves with irregularly voulted surface and downwards pointing tips), chlorosis and necrosis, epinasty and ultimately no symptomss at all if they are infected at full maturity. Chlorosis and vein necrosis occur only when leaves are infected during the growth stage in which cell elongation predominates over cell division.

Such a difference in reaction to virus infection may often be observed when comparing apical and basal parts of a single leaf, because leaves mature earlier at their apex than at the base. Thus of inoculated plants the youngest leaves treated may show local lesions near the apex and vein chlorosis or other systemic symptoms near the base (e.g. Solberg & Bald, 1962), or the first leaves producing systemic symptoms may show them in the basal part only (Fig. 5 right).

Another reason for the sequential production of many symptoms is that often one abnormality causes the other. So the tomato black ring virus incites top necrosis in *Chenopodium amaranticolor*, in turn causing loss of apical dominance and often excessive branching of the stem (Fig. 7). The next example is more complicated. Virus translocation through the vascular system may cause rapid necrosis there. This may disturb transport, causing starch to accumulate and altering the texture and colour of the entire leaf or parts of it. Obstruction of water supply causes wilting and other phenomena. The final consequences go far beyond the original effect of the virus on host biochemistry.

During a disease, symptoms often pile up or progress in severity (progressive debilitation or progressive deterioration). This effect is known as *decline*, either gradual (slow decline) or rapid (quick decline). It is common in fruit tree virus diseases and has been studied especially for citrus tristeza or citrus quick decline (Schneider, 1954; for a survey, see Schneider, 1959). This disease is known only for certain scion-stock combinations, especially of sweet orange on sour orange. Here the syndrome starts with phloem necrosis in the highly sensitive sour orange rootstock immediately below the bud union. Roots then become depleted of starch and eventually rot. Hence, sweet orange aerial parts of the tree, that are tolerant to virus multiplication as such, are progressively starved and show wilting, defoliation and associated phenomena (see also phloem degeneration, p. 53).

Recently this syndrome has received considerable attention in pear

Fig. 7 Bushy growth due to apical necrosis (right) in *Chenopodium amaranticolor* with tomato black ring virus. (After Bos, 1965b.)

decline, a disease of great economic importance now ascribed to mycoplasma-like micro-organisms (see also p. 151). Infection generally manifests itself in commercial varieties of *Pyrus communis* budded or top-grafted on *P. serotina* and *P. ussuriensis* rootstocks. Necrosis and degeneration in sieve tubes below the bud union may girdle the tree. In subsequent slow decline, growth is stunted and leaves are small and pale green. In quick decline, apparently normal trees and slowly declining specimens suddenly wilt and the leaves turn brown. Infected trees that look normal and those with slow-decline symptoms may develop red foliage early in autumn, often together with a brown discoloration of the veins in mature leaves. Trees with slow decline usually linger indefinitely with a gradual loss of yield and vigour (Batjer & Schneider, 1960; Griggs *et al.*, 1968).

Plants weakened by virus infection become more susceptible and sensitive to other adversities or to other pathogens and pass through a complex syndrome (p. 117).

Finally the term *degeneration* should be mentioned. It refers to a gradual decline in health of vegetatively propagated varieties of potatoes and other plants. During continuous cultivation such varieties may become completely infected with a certain virus and often with an increasing number of viruses. The resulting progressive reduction in vitality, or 'running out' of a potato variety has long been known (e.g. Atanasoff, 1922b); until the beginning of this century it was ascribed to mere ageing and to the unnaturalness of continuous asexual reproduction.

Plants may fight back

In virus-infected plants, symptoms do not always increase in severity. An acute phase is frequently followed by a chronic phase and symptoms may even be completely absent from later formed organs. Both from a practical and a theoretical standpoint this phenomenon, especially its physiology, is of utmost importance. In recent years it has attracted much attention and its study has shed new light on the relationships between host and virus, indicating that the host is not just a passive subject to virus infection but that it often actively 'fights back'. It therefore deserves fuller discussion here, as already done in Dutch (Bos, 1965a).

The *acute phase* may occur soon after inoculation; it may often be shock-like and is then called the *shock phase*. It is characterized by severe symptoms that may even lead to the host's death. If the host survives, a *chronic phase* commonly follows, often characterized by some *recovery*, meaning that in newly formed parts symptoms are less severe. The plant may even recover completely from disease (Fig. 8). Although symptoms may completely disappear, recovered plants still contain the virus. There is only one report on the complete disappearance of virus from recovered plants viz. by Ivancheva-Gabrowska (1974) on tomato spotted wilt virus from tobacco.

Wingard (1928) first described this phenomenon in tobacco plants infected with tobacco ringspot virus. Later it was observed in various diseases and proved typical of all ringspot viruses. Price (1932) checked Wingard's findings and showed that the virus was still present in plants after ten generations of vegetative propagation. However, virus concentration was greatly reduced in recovered parts of the plant and later Price (1936) found them to contain only a sixth to a tenth of the amount in severely diseased parts. In general it seems that in a certain plant or variety there is a correlation between symptom severity and virus concentration. This view was supported by work of Ross (1941) with alfalfa mosaic virus in tobacco plants. Infected plants did not completely recover, but the chronic phase was very mild. Ross mentioned a striking correlation between symptom disappearance and reduction in virus concentration. But in tobacco plants with tobacco mosaic virus no such reduction in virus concentration could be observed (Steere, 1952), so that the tobacco mosaic syndrome remained constant.

Wallace (1944), studying tobacco and tomato plants infected with beet curly top virus, was the first to demonstrate that plant defence has a material basis. Insects easily transmit the disease, whereas tomato plants infected after grafting from tobacco show only very mild symptoms. The latter has been ascribed to the presence of graft-transmissible 'protective substances' in recovered tobacco plants. Evidently, the plant possesses some kind of defence mechanism, triggered only by infection; it may actively resist. The term 'acquired resistance', introduced by Wallace (1944) is sometimes used for this.

Research on this 'mechanism' gained new impetus after an interesting discovery by Ross (1961a). He found that leaves of 'Samsun NN' tobacco locally reacted to tobacco mosaic virus with discrete necrotic local lesions and that another inoculation of the same leaves a few days later again led to the production of local lesions. But one or two millimetre wide zones surrounding the first-formed lesions remained free or contained at most a few tiny ones. Thus, the localization of infection by tobacco mosaic virus in this host does not seem to be primarily due to hypersensitivity, producing a mechanical barrier of rapidly dying tissue, but mainly to some substance diffusing from the infected cells. Such a localization of infection may also occur in virus-host combinations reacting with chlorotic or yellow lesions or with no visible local effect at all.

Soon afterwards, Ross (1961b) discovered that acquired resistance could become truly systemic. When the lower half of a leaf was first inoculated and the upper half 2-20 days later, or vice versa, or the right half first and the left second, then the lesions from the second inoculation were much smaller than from the first inoculation. This systemic effect could also be observed in top leaves of a plant of which only the lower leaves had been inoculated. The uninoculated resistant

Fig. 8 Recovery after a shock phase in 'White Burley' tobacco sap-inoculated with alfalfa mosaic virus. The three top leaves are symptomless.

leaves or half-leaves always proved to be virus-free. The resistance induced by tobacco mosaic virus held also for tobacco necrosis virus, for turnip mosaic virus and for tomato and tobacco ringspot viruses. Later this phenomenon was also observed by others with other viruses and other plants. The existence has been postulated of substances resembling interferon, which was discovered by Isaacs & Lindenmann (1957) to play an important role in animals and man, in addition to more specific antibodies interfering with virus infections. So, protective substances produced as a result of virus infection in plants seem to limit the size of lesions and to stimulate recovery. Tentative research on an antiviral factor from virus-infected plants and its characterization as a protein of 20 000 to 30 000 daltons in molecular weight (e.g. Sela *et al.*, 1966), has not yet led to a better understanding of the mechanisms of acquired resistance and local-lesion formation.

Strangely, after a chronic phase or apparent recovery from disease, another acute phase may ensue, and so on. The first to draw attention to this phenomenon were Cheo & Pound (1952) in their study on tobacco plants infected with cucumber mosaic virus. They observed a coincidence between the syndrome and a cyclic course in the virus concentration of their plants. Paul & Quantz (1959) associated such a cyclic inhibition of true mosaic virus disease in broad bean with the possible production of an 'inhibiting' substance. It takes time to be produced: first the virus gains its chance; then protective substances diminish virus multiplication; less of these substances is produced because of their redundance; the virus gets another chance, and so on.

Paul & Quantz based their explanation also on experiments with the 'Echte Ackerbohnenmosaik-Virus' (broad bean true mosaic virus) in broad bean plants. Later on Wasuwat & Walker (1961a, b) and Marrou *et al.* (1968) confirmed the results with cucumber mosaic virus in pumpkin and cucumber, and Selman & Milne (1961) with tomato spotted wilt virus in tomato plants. In broad bean and tomato, infected plants attracted attention by alternating 'storeys' of diseased and healthy looking leaves. I have observed a similar periodic shift in symptom severity in pea plants infected with cowpea mosaic virus and sometimes with pea enation mosaic virus. In no case was temporary symptom disappearance due to temporary change in conditions.

Such an intermittant or undulating repression and stimulation of virus multiplication and associated symptoms may also underly concentric ringspotting often occurring at the site of inoculation or sometimes after systemic infection (Bos, 1965a; Fig. 9). In line pattern and oak-leaf pattern a similar sequence of severely and mildly reacting zones of tissue is often observed. In extracts of infected tobacco leaves, a rhythmic course of the concentration of cucumber mosaic virus with periods of some thirty hours was indeed detected (Marrou & Migliori, 1968).

Absence of symptoms

As already said, virus multiplication in susceptible hosts does not always lead to visible symptoms. This absence of perceptible deviations is known as *inapparency*.

Nishimura (1918) has found that tobacco mosaic virus produces such a symptomless infection in *Physalis alkekengi* and he is the first

Plate 1 Colour breaking in flower of Triumph tulip 'Elmus' with tulip breaking virus. (From Plant Protection Service, Wageningen.)

Plate 2 Local lesions by a strain of bean common mosaic virus in primary leaves of *Phaseolus vulgaris* 'Monroe', 13 days after inoculation. The lesions extend along the veins.

Fig. 9 Unusual concentric ringspotting in 'White Burley' tobacco, presumably with raspberry ringspot virus. (From Miss H. J. Pfaeltzer, Wageningen.)

to draw attention to the phenomenon. A few years later James Johnson (1925) and Atanasoff (1926) detected viruses in apparently healthy potato plants. Since then the problem of inapparent infection has received considerable attention, especially in the certification of virus-free propagation material. But inapparency is a fairly general problem in the control of virus diseases, since symptomless hosts may act as a source of infection for neighbouring sensitive plants of the same stock, other varieties or other crops. Such symptomless plants are usually called 'symptomless carriers' or, for short, 'carriers'. This is especially important since we know that viruses usually are much more 'polyphagous' than long supposed, and many viruses that are harmful to cultivated crops are 'carried' unnoticed in natural vegetation. A review of older literature, a great many examples, and practical implications of inapparent infection are given by Hildebrand (1958). An illustrative example in fruit trees has been mentioned by Kegler (1964a): 2.1 to 34.6% of the investigated trees showed symptoms (2.1% for sweet cherry to 34.6% for sour cherry), while positive tests for virus infection ranged from 27.2% in sweet cherry trees to 95.5% in apple trees. Latent infection has proved widespread in apple, sweet cherry, and plum.

Whether symptoms are produced or not depends on the virus, the host species and external conditions. Some aspects will be discussed more in detail.

Of course, even in the most sensitive host there always is a symptomless period immediately after inoculation and the onset of in-infection. This presymptom period or *incubation period* may vary in length from some days or even shorter to over a year, again depending on the virus, the host species, and conditions. The term 'latency period' is sometimes used for this or the period after virus uptake by a vector before it is able to transmit the virus.

Many viruses have hosts in which infection never causes visible symptoms. In plants this permanent inapparency is called *latency* (L. latere = to lie hidden). Such hosts lack *sensitivity*; they show extreme *tolerance*. But the presence of virus can be demonstrated by back-inoculation to sensitive hosts, by serology or by electron microscopy. Some new viruses even have been discovered in sap from symptomless plants inducing symptoms in experimental hosts; examples are dodder latent virus (Bennett, 1944) and carnation latent virus (Kassanis, 1954, 1956).

Virus infection in wild plants often lacks visible effects, perhaps because of natural selection for latency. Presumably such viruses have been associated with these plants for long periods. Probably the viruses have reached a state in which they can persist by causing a minimum of injury to their host (Bennett, 1958). Thus the genetics of plant and virus may have adapted to each other: plants suffering from infection, and hence genetically sensitive to virus, and plants infected with virulent virus strains, have been eliminated in the struggle for life. Man has promoted latency in various cultivated plants, especially in vegetatively propagated ones, by ignorantly selecting for tolerance. Latency may have been further encouraged by the easy-going usage of the term resistance, as selectionists and breeders look mainly for visible effects without bothering about the role of the virus itself. As with cucumber green mottle mosaic in cucumber (Čech & Branišová, 1976), tolerance may be associated with genetic differences in concentration of virus. Of course, low concentrations are preferable from an ecological point of view.

In a discussion on some aspects of 'gradations between pathogenicity and commensalism in infections with plant viruses', Bawden (1958) has proposed to use the term 'commensalism', known from other fields of biology, for latency with viruses. There is no direct need for such a term, but it neatly describes the frequent harmonious coexistence of plant and virus.

Symptoms may also disappear temporarily. Newly formed organs of virus-diseased plants are then free of symptoms, whereas in those formed some time later abnormalities may turn up again. This temporary absence of symptoms is commonly called *masking*. It is often caused by environmental factors, such as light and temperature. A striking example is the complete disappearance in summer of a bright vein yellowing (Fig. 18) from lucerne plants infected with bean leafroll virus (van der Want & Bos, 1959). A good illustration is the influence of temperature on the symptoms of prune dwarf in Italian prune (Moore, 1967). This disease usually remains masked at temperatures above 13°C (Fig. 10) but with day temperatures of 22°C it develops, as long as nights are cool (at least if the temperature on one night out of every three is 13°C or less). Virus however, is as readily transmitted from abnormal as from healthy-looking parts. With certain fruit-tree virus diseases the symptoms can only be observed at irregular intervals, without known relationship to environmental conditions. This phenomenon is sometimes called *recurrency*.

If the disease becomes almost permanently masked, this may be called recovery, as discussed earlier (p. 27). Physiologically this is something fundamentally different. Total recovery and masking can only be distinguished from latency by the presence of symptoms in older parts of the plant.

In pathology much confusion exists on the definition of latency and masking as two forms of inapparent infection. According to Gäumann (1946) 'inapparent' means permanent absence of symptoms, and 'latent' their temporary absence. Thus in cereal and grass smuts, Butler & Jones (1949) speak of a latent infection, showing up only in the grains at the end of the life of the host. Since latent literally means hidden or dormant, this usage is accurate; as is the expression latency period (formerly called latent period) of the virus in its vector.

At a medical 'symposium on latency and masking in viral and rickettsial infection' (1958) it was suggested that the term 'inapparent' be used for all infections without visible symptoms, the term 'latent' only for those inapparent chronic infections that result from a balance between host and virus, and that the term 'masking' be dropped. Indeed 'masking' literally means 'covered to disguise', 'unrecognizable but still visible', and its application in the present sense seems incorrect. But since it is currently used in plant pathology (e.g. Klinkowski *et al.*, 1967), and there is a need to distinguish between permanent and temporary absence of symptoms, it should be maintained alongside apparency and latency, despite its literal meaning.

We now come to the problem whether inapparency really exists. Theoretically a truly latent infection may be possible, e.g. if virus could multiply on material not required for normal cell metabolism. Though as yet no exact data are available, it is known that virus multiplication requires exactly the same building blocks as those essential in normal physiological processes of plants, so that viruses cannot grow without some influence, though it may be undetectably slight.

This brings us back to the problem outlined in defining the *disease* concept. It was then concluded that there is no sharp limit between normal and abnormal or between 'sick' and 'healthy' in plant growth because of the natural variation in normal development (e.g. Küster, 1925; Bloch, 1954).

Whether symptoms are perceptible or not, and whether infection is apparent or inapparent depends largely on the accuracy of the observations. The electron microscope and a careful study of physiology and biochemistry of infected plants should reveal many changes due

Fig. 10 Masking of symptoms in Italian prune with prune dwarf virus above 13 °C. Continuous cold (A); temperature at first below, later above 13 °C (B); the reverse of B (C); continuous warmth (D). (From Dr J. D. Moore, Madison, Wisc.)

to viruses that have still escaped observation.

How difficult it may be to detect a disease has been demonstrated in potato virus research. Atanasoff (1926) found that grafting ostensibly healthy potato plants 'Zeeuwse Blauwe' on 'Eersteling' ('Duke of York') led to severe symptoms of stipple streak (virus Y) in the latter. James Johnson (1925) studied this phenomenon in more detail and found that extracts from potato plants, even if seemingly healthy, always produced virus diseases when grafted onto tobacco and some other solanaceous plants. This capacity seemed to be present in most, if not all, standard potato varieties.

A more recent example is potato virus S, discovered by de Bruyn Ouboter (1952) after checking incongruities met with in serological research on other potato viruses. Later studies by Rozendaal & Brust (1955) showed a very high incidence of virus and almost complete infection in several potato varieties. A few varieties produced a slight mosaic, whereas most developed symptoms perceptible only by careful comparison of infected and virus-free plants. In comparable experiments by Scholz (1962), the virus caused an average decrease in yield of 12% in four varieties tested under different conditions.

Carnation latent virus (Kassanis, 1954, 1956) and *Passiflora* latent virus (Schnepf & Brandes, 1961; Brandes & Wetter, 1963), distantly related to potato virus S, gave symptomless infections in their main hosts. 'Ladino' white clover often decreased in yield after infection with a mixture of bean yellow mosaic virus and alfalfa mosaic virus, even when symptoms were hardly visible (Kreitlow *et al.*, 1957).

Obviously the separation of apparent and inapparent infection is rather arbitrary; it often depends on the accuracy of the technique.

Variation of symptoms

In the examples so far, and in those still to be described, the reader will notice a great diversity of symptoms, because of the vast numbers of viruses and strains. *Datura metel*, recommended as a virus-indicator, consistently reacted with 28 distinct symptoms to 19 isolates of 12 viruses (Kahn & Monroe, 1970). There are also differences in reaction of various hosts to one virus. In addition, a virus may manifest itself in various ways according to external conditions and stage of development of the host or its parts.

The influence of such conditions on symptom expression is easily understood since virus disease is almost exclusively a reaction of host metabolism. Even in healthy plants metabolic processes vary considerably with environmental conditions, but they remain within the limits set by the cell's genetic control mechanism. Virus-infected plants largely or completely escape that control.

The age of the attacked tissue or organ is an important factor in determining host reaction. One of the many examples has been supplied by Bos & Maat (1965) in their study on red clover mottle virus in leaves of 'Beka' beans, showing a great number of chlorotic local lesions if inoculated ten days after sowing, whereas if inoculated six days later numerous tiny necrotic lesions develop. Likewise, upper and lower leaves of *Chenopodium* plants may differ considerably in their type of local reaction to several viruses.

When a plant is systemically invaded by a virus, old and young tissues are infected at about the same moment and they may react in completely different ways, as mentioned for bean common mosaic virus (p. 24). Similarly, systemic host reaction may vary considerably in plants infected at different stages of development: Ford (1964) has provided evidence that in *Pisum sativum* infection with pea streak virus leads to necrosis in the pods only when the plants are infected at a particular late stage of development.

As light is essential for plant growth, it also influences virus symptoms. For many viruses the symptoms in greenhouses are weaker in summer than earlier or later, or in artificial shade, perhaps because plants grown with less light contain more virus. Size of lesions, and extent of necrosis and pigmentation in the lesions caused by alfalfa mosaic virus in leaves of French bean are smaller when the plants are kept almost dark for some time after inoculation (Desjardins, 1969). The effect of light is partly explained by its influence on host susceptibility and virus concentration (Opel, 1970).

With bright light in summer, symptoms may completely disappear, as in vein yellowing of lucerne caused by bean leafroll virus (van der Want & Bos, 1959). Conversely, high light intensity was found essential in California for symptom expression of yellowing diseases in sugar beet, lettuce and spinach among others caused by beet western yellows virus. Portions of the fields in the shade of giant eucalyptus rows remain green whereas the rest of the field may be golden yellow (Duffus, 1977).

Casper (1968) suggested that the composition of the light may also play a role. He stated that blue light promotes virus multiplication and the appearance of symptoms in *Nicotiana* infected by *Arabis* mosaic virus and potato virus X.

Temperature, of course, often determines the nature and severity of symptoms because it strongly influences virus concentration (for a survey see Wolffgang, 1970a), rapidity and course of infection and host sensitivity. It may be important in determining whether or not local lesions are formed and in defining the moment lesions appear, their number, and their size (e.g. Wolffgang, 1970b). But there is no general rule. A shift in temperature may induce a striking change in syndrome. Hypersensitive varieties of French bean, normally reacting only locally to bean common mosaic virus, are usually infected systemically at temperatures above 20 °C and then show severe vascular necrosis known as black root disease (Grogan & Walker, 1948). The same holds for tobacco mosaic virus in *Nicotiana glutinosa* which is hypersensitive at normal temperatures but is systemically infected at 35 °C or more (Samuel, 1931).

Another example of the impact of temperature is gherkins, which up to maximum day temperatures of 25 °C normally develop mosaic after infection with cucumber mosaic virus, but wilt if the weather is cloudy cool (with temperatures of 20 °C or below during the second week after inoculation); necrosis and mottling develop at 20° to 25 °C (Tjallingii, 1952). Later similar wilting was found by Schmelzer (1967) for pumpkin and squash after low temperatures during the incubation period of this virus, whereas for cucumber green mottle mosaic virus in cucumber Wache (1966) demonstrated a clear dependence on temperature and air humidity of symptoms in inoculated and in systemically infected leaves.

A study of 21 virus diseases in 24 deciduous fruit trees showed that with most virus-host combinations the severest symptoms occurred at 18 or 22 °C. Others had most severe symptoms at 26 °C. Some combinations had symptoms at all four temperatures tested and others only within a restricted range of temperatures (Fridlund, 1970). At certain temperatures symptoms may even be lacking or disappear, as in prune dwarf (Fig. 10). For some further examples of the influence of changing temperature on plant virus diseases, see Kassanis (1957).

Parts of several plant species, especially woody ones such as fruit trees and roses, have even been freed from virus infection with artificially raised temperatures nearly lethal to the plants. This technique produces virus-free scions or budwood (e.g. Nyland & Goheen, 1969).

Of course, nutrition also influences host reaction. Nitrogen dressings may have such a side-effect by turning leaves dark-green, so covering or masking leaf variegation. With tobacco necrosis virus in French beans, Hofferek & Wolffgang (1964) found that application of nitrogen considerably reduces symptom severity, without decreasing virus concentration. Hence, excessive rates of nitrogen should be avoided when growing seed potatoes.

Hence, for optimal symptom expression when working with

indicator plants, e.g. for virus detection, manipulation of the experimental environment is of utmost importance (Bos, 1967, 1976). Conditions have to be standardized when reaction of test plants is used for routine indexing.

In nature multiple infections by different viruses or by viruses and other pathogens often occur, each pathogen on its own influencing host physiology and thus altering host susceptibility, sensitivity or resistance to the other pathogens, either diminishing or aggravating their effects (*antagonism* or *synergism*). Hence, the resulting syndromes may differ completely from the effects of each pathogen alone. See also the section on increased predisposition to non-virus diseases.

Theoretically, the effect of certain virulent viruses could be alleviated by selection of growing conditions or even inoculation with certain virus, virus strains or other pathogens, but under field conditions the various factors are often hard to manipulate.

Description of symptoms

To be readable yet suitable for reference, a review of virus symptoms should start with some kind of classification, to allow a clear definition of names and terms.

The starting point of this chapter will be our present knowledge of pathogenesis as set out in the previous chapter. This knowledge is based on virus structure and biochemistry (which already predict various aspects of virus-host relations) and on host reactions (further elucidating these virus-host relations).

Because of the nature of viruses, diseases start with *biochemical and metabolic changes*, though these may also result secondarily from structural deviations in the host. Most of these changes are hard to detect and usually have no diagnostic value.

After the immediate disturbance of cell metabolism, symptoms usually initiate inside the cells and tissues of the plant as *anatomical deviations*. These finally result in macroscopically visible symptoms, which can be grouped into two main categories. The first comprises all aberrations more directly resulting from abnormal cells and tissues; the second those which are due to abnormal development of tissues that consist of elements that themselves may remain normal. The first category includes such abnormalities as grouped in the sections on *growth reduction*, *colour deviations*, *water deficiency* and *necrosis*; the second those under *abnormal cork formation* and *malformations*. The deviations that cannot be brought under one of these headings will be treated as *miscellaneous abnormalities*.

There are cases in which abnormalities are not directly due to, though associated with virus infection; they are caused by other factors introduced by previous virus infection: *increased predisposition to non-viral diseases*.

Of course this classification is only for the sake of arrangement and cannot always be strictly adhered to.

Recent studies have supplied fascinating data on cytological deviations. Because extensive studies by electron microscopy have accumulated data on the physical participation of the viruses themselves, the chapter closes with a section *virus particles and inclusion bodies*.

Biochemical and metabolic changes

It is now accepted that plant viruses incite disease in a manner completely different from that of pathogenic organisms, besides entering their hosts entirely passively. The detrimental effects of bacteria and fungi on the hosts are mainly due to depletion of host metabolites by the parasite or to toxic metabolic products of the parasite. Anyway, these organisms take part in the final syndromes as visible bodies. Until recently, virus particles completely escaped observation in virus syndromes. Most viruses seem to lack any metabolic system of their own. Finally, nucleic acid isolated from purified virus suspensions has been shown to initiate infection completely on its own. Thus, metabolic products of the virus itself are completely ruled out.

With virus diseases pathogenesis has to be considered a genetic process where, from the moment of infection, metabolism of the host is no longer entirely governed by the host's genetic system alone. Genetic information encoded in the virus's nucleic acid interferes, leading to physiological and, more specifically, biochemical disturbances.

Much work has been done on such physiological and biochemical changes in virus-infected plants to trace the initial events caused by infection and for a better understanding of pathogenesis. Of course, many of these investigations have also been carried out to find possible ways of interfering chemically with infection. A considerable part of the work has been devoted to the search for reliable differences between healthy and diseased plants to support diagnosis and even to allow diagnosis in symptomless infection.

Various authors have reviewed the physiology and biochemistry of virus diseases and virus infection (Bawden, 1959; Porter, 1959; Diener, 1963; Bawden, 1964; Hofferek & Wolffgang, 1964; Köhler, 1964; Farkas & Solymosy, 1965; Opel, 1965; Matthews, 1970).

Many changes have been reported in virus-infected plants, varying with the virus itself and the host. They show that the effects are extremely variable, as in normal metabolism, and depend closely on leaf age, diurnal and seasonal variations and on external growth conditions. The difficulties in studying them should not be underestimated as shown, for example, by Opel's work (1965) with homogenized leaf tissues of which only part is affected.

The more important biochemical and physiological aberrations caused by viruses will be summarized now; they are taken mainly from the above surveys, without reference to the original literature.

In sap of virus-diseased potato tubers the redox potential has been reported to be lower than normal, electric conductivity higher and freezing point lower. In sap of such tubers, and in leaves of potatoes and some other plants, buffer capacity has sometimes increased. A few reports suggest the production of special toxins or mention changes in alkaloid content. Often organic acid content is higher in virus-infected plants.

In many cases virus infection has considerably reduced photosynthesis. This can only partly be ascribed to the reduced chlorophyll content caused by many viruses and seems to be greatly influenced by enzymic disturbances. Respiration often increases in virus-infected tissues, but has sometimes been observed to decrease. Of course, these abnormalities largely depend on shifts in enzymic activities. But many reports are contradictory.

Increased polyphenol oxidase activity has attracted special attention, mainly in connexion with the production of necrotic local lesions. Polymerization of phenolic compounds, such as the chlorogenic acid, resulting in polyphenol derivatives like melanins and quinones, leads to a shift in light absorption from ultraviolet to visible wavelengths, hence the violet and black colours. The quinones are supposed to be toxic and to induce necrosis and hypersensitivity reactions, but their role is not completely understood. These substances are also known to accumulate during senescence of certain tissues and, in virus disease, they may merely result from physiological disturbance caused by the virus (Diener, 1963), especially at death, as their enzyme-controlled production may continue after cells are killed, as seen for instance, in expressed sap turning black.

Linked with disturbed photosynthesis, but also with disturbed enzyme-controlled transport, are deviations in starch content. *Accumulations of starch* can readily be demonstrated by staining with iodine potassium iodide after decolorizing the leaf with ethanol (Holmes, 1931). Thus, leaves harvested after an extended period of darkness may show dark lesions where starch deposits remain because of restricted translocation (Fig. 11). In contrast, after a long period of light, paler lesions may stand out against a dark background owing to restricted synthesis of starch in these lesions. Large amounts of starch and other carbohydrates can accumulate in leaves of potato infected with potato leafroll virus (e.g. Murphy, 1923) and of

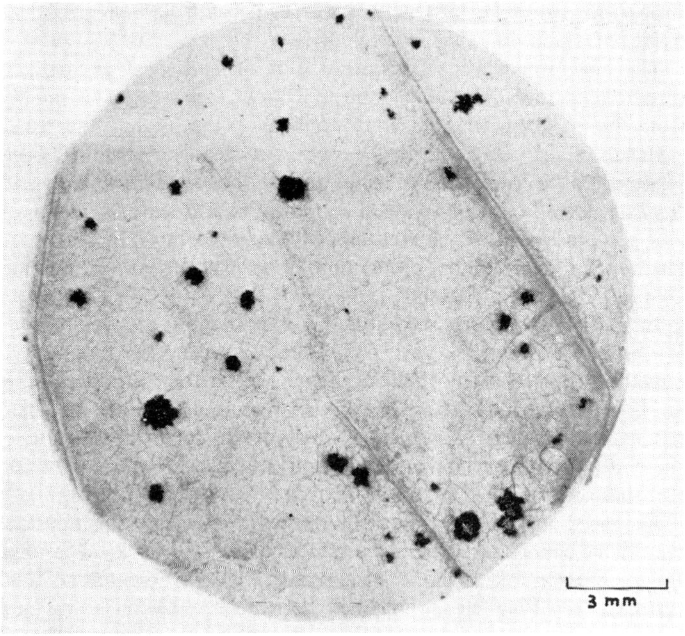

Fig. 11 Starch lesions in a bleached iodine-stained leaf disc of 'Pinto' bean *(Phaseolus vulgaris)* 4 days after inoculation with Ul strain of tobacco mosaic virus (× 4.5). (After Helms & McIntyre, 1962.)

sugar-beet with beet yellows virus (Watson & Watson, 1953).

For a long time reduced transport of starch, especially in potato, was assumed to be directly due to collapse and necrosis of sieve tubes (Quanjer, 1913). But later starch accumulation was found to occur before phloem necrosis became detectable (Murphy, 1923), and the severity of the leaf symptoms does not seem to be correlated with the degree of phloem necrosis (Wilson, 1955). Work of Henke (1957) on sugar-beet yellows suggests that transport of carbohydrates is primarily disturbed by abnormal activity of phosphatase.

Much effort has been spent in analysing nitrogen metabolism, with special reference to proteins and amino acids, because accumulation of virus-specific abnormal proteins can be enormous. In mosaic-diseased tobacco plants the total virus content has been reported to reach 11.7% of total dry weight (Steere, 1952), or 50% of all the protein. Virus protein may be formed in excess of what is needed for coating the virus nucleic acid, as for tobacco mosaic virus (Newmark & Fraser, 1956) where it has been called X protein (Takahashi & Ishii, 1953).

In addition to such mainly coat protein, a range of other abnormal non-coat proteins have recently been shown to occur in tobacco infected with various viruses. In 'Samsun NN' tobacco with tobacco mosaic virus up to twelve different soluble proteins with molecular weights mainly between 10 000 and 20 000 daltons have been demonstrated by electrophoresis in polyacrylamide gel (van Loon, 1972). Interestingly, similar proteins were detected in viral-like lesions in 'Samsun NN' tobacco induced by local administrations of the ethylene-releasing compound 2-chloroethylphosphonic acid (van Loon, 1977). These proteins differ from those obtained by purifying and breaking down intranuclear crystals isolated from tobacco etch virus-infected plants and cylindrical components from amorphous inclusions from plants with a number of potyviruses and to be discussed further under inclusion bodies. The latter proteins are ca. 50 000 to 100 000 and ca. 70 000 daltons in mass, respectively, are virus-specific and differ antigenically from host protein.

Only a little work has been done on the effect of virus infection on growth-regulating substances (see also malformations, p. 87). Hormonal disturbances are evident: in various cases a virus-induced reduction in the amount of growth-promoting substances or an increase in concentration of growth-inhibiting substances have been suggested. A growth substance, ethylene, may even play an intermediate role in local-lesion formation and acquired systemic resistance (van Loon, 1977).

The production of interferon-like substances after infection has already been discussed in the previous chapter when dealing with local and systemic acquired resistance and recovery from disease.

Some biochemical changes are directly visible, such as a decrease in chlorophyll content leading to chlorosis and yellowing (further enhanced by an increase in the concentration of carotenes and xanthophylls) and an increase of anthocyanins resulting in reddening (see under colour deviations). The same holds for the accumulation of dark-coloured melanins.

Thus, from purely physiological and biochemical studies of the infected hosts, the nature of changes in metabolism of virus-diseased plants has not yet been clarified. Various abnormalities point to a non-specific but comprehensive derangement of normal host metabolism. How this is brought about is unknown. There is no proof that any virus attacks any specific cell component. Authors are gradually becoming convinced that virus multiplication irreversibly deprives cell metabolism of building blocks essential for the cell's normal chemistry and that most derangements studied are purely secondary phenomena, late occurrences, not indicative of the initial events directly resulting from infection. Admittedly, this theory seems conflicting with the many latent infections often associated with high virus concentrations. Usually clear metabolic changes are not produced until clear-cut symptoms have been observed. Commonly this is also the time of maximum virus multiplication.

The conclusion is gradually gaining ground that many virus symptoms must be explained as premature ageing or senescence, or as shortage of essential building blocks or energy-supplying substances. Accumulation of toxic substances that are dissociated in healthy plants, or that are especially produced by intensive virus multiplication, might also result. Many normal processes, such as the breakdown of proteins and chlorophyll, are accelerated after virus infection.

Research on *Phaseolus vulgaris* with tobacco necrosis virus (Hofferek, 1967a, b; see also Hofferek & Wolffgang, 1964) have shown that various effects of infection, such as symptom expression, and certain changes in enzyme activity can be partly or completely overcome by the application of phosphate or ammonium nitrate and ammonium sulphate, or all three, whereas virus concentration does not decrease or sometimes even increases. In some plant species virus-induced systemic chlorosis and chlorosis around necrotic lesions as well as necrosis are similarly inhibited by decapitation, by treatment of detached leaves with ammonium nitrate or with kinetin (Opel, 1965), or by rooting of detached leaves (Opel, 1967), all rejuvenating the plants or tissues concerned.

Thus, in spite of considerable differences in initiation, many symptoms of virus diseases closely resemble the effects of other pathogens, or especially physiological disorders such as deficiency diseases, or genetic abnormalities (see also next chapter, p. 144). Characteristic, however, remain the way, the distribution, or the sequence of occurrence in virus-infected plants because of the typical way the virus infects and spreads inside the plant, or because of the special relationships between certain viruses and certain tissues.

These special relationships already partly answer the important question whether certain types of symptoms can be characteristic of certain viruses (and thus can be used for diagnostic purposes). Which virus proteins are made, and thus which host amino acids are removed and to what extent this occurs, all depends on 'instructions' encoded by the nucleotide sequence in the virus nucleic acid. Of course, the same holds for the incorporation of nucleotides into virus nucleic acid instead of host nucleic acids. This 'parasitic' virus multiplication determines which energy processes, building processes or enzymic processes of the host will get out of balance and which, finally, via a chain of associated processes, will be the externally visible effect on the host.

A related question is whether the virus genome contains special information influencing host reactions. In tobacco mosaic virus only a part of the total genetic information is thought to be needed for coding coat protein synthesis and the production of the special polymerase. Strains of this virus are known that are indistinguishable in amino acid sequence, nucleotide composition, electrophoretic behaviour, isoelectric point, and serological affinities (literature reviewed in Ragetli, 1967), but that clearly differ in symptoms. In this respect the before-mentioned virus-coded non-coat proteins increasingly attract attention. They are likely to be coded for by virus genes other than those for virus multiplication and the production of coat protein and may influence symptom expression. That a virus may carry genetic information influencing the host but not the virus itself is difficult to understand on evolutionary grounds.

So in virus pathogenesis only a few problems have been solved and many more questions remain to be answered. There is little sense in further discussing biochemistry and physiology of virus infections here because, as remarked above, the deviations they cause can hardly be considered symptoms. Deviations leading more directly to visible effects or to stainable host reactions will be discussed in relevant sections, especially in the next one (see also Lindner, 1961). Several techniques have been developed to diagnose virus infections in plants rather than to study pathogenesis; few have become established and still fewer are generally accepted. Here again we come up against the lack of basic differences from normal but rather there are a series

Plate 3 Yellowing in outer leaves of head lettuce with beet western yellows virus; right, healthy plant.

Plate 4 Leaf chlorosis and reddening in carrot with carrot red-leaf virus.

Plate 5 Leek yellow stripe caused by a virus related to onion yellow dwarf virus.

Plate 6 Rose leaves with yellow line patterns and bands ('rose mosaic') after infection by *Prunus* necrotic ringspot virus.

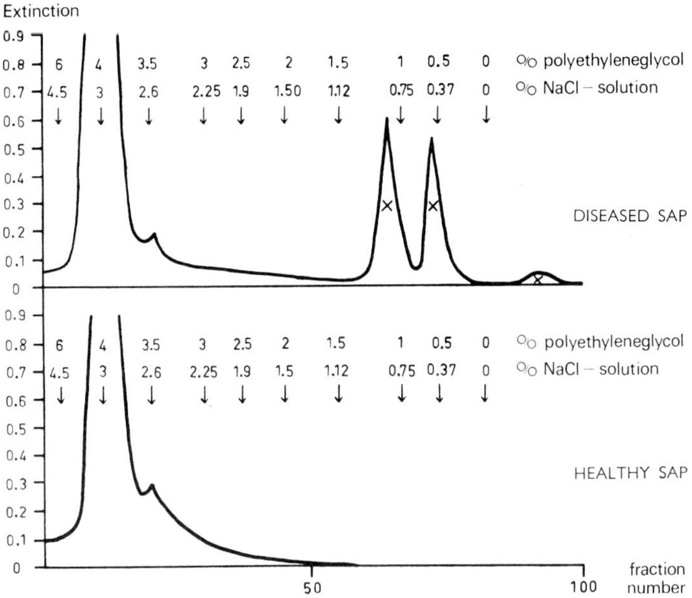

Fig. 12 Difference in ultraviolet absorption at 254 nm between sap from tomato plants infected with potato virus X (top curve) and from healthy plants (bottom curve) chromatographed through a cellulose column. The aqueous eluents contained different proportions of polyethylene glycol and NaCl as indicated by the arrows. Presence of infectious virus is indicated by X signs. (After Venekamp & Mosch, 1963.)

of grades. When populations of plants are tested, certain chemical properties show wide distribution curves overlapping for normal and diseased groups, so that an individual specimen is hard to place. Thus such tests are often unreliable (compare excessive formation of callose in sieve tubes, p. 49).

The best established biochemical difference between healthy and virus-infected individuals is the presence of virus nucleoprotein. The viral proteins, and especially the viral nucleic acids differ from their host's counterparts in the sequence of their building blocks. They can be distinguished from normal host components only by physical structure and size as revealed by electron microscopy, ultracentrifuging, electrophoresis or chromatography (Fig. 12) (as often facilitated by their abundance resulting from almost unlimited multiplication), and by their infectivity in tests with indicator plants.

As yet there are no simple and reliable chemical routine tests for the virus nucleoproteins themselves. But results with electrophoresis on sodium dodecyl sulphate polyacrylamide gels to rapidly detect a number of viruses by their specific proteins (Paul, 1974, 1975) seem promising.

Anatomical deviations

Since viruses act via their host's cells to cause externally visible effects, morbid anatomy has constituted an integral part of pathogenetic study ever since virus diseases of plants were first studied (e.g. Iwanowski, 1903; Quanjer, 1913). Later it gained attention from those concerned with diagnosis of virus diseases because, like biochemical changes, anatomical derangements may occur without external effects, be visible earlier or be more characteristic than the external changes.

To study the anatomy of infected plants, a wide range of techniques is available.

Conventional cytology has been greatly assisted by the use of dyes and the phase-contrast microscope. Histochemical (or better cytochemical) techniques trace chemical changes, together with micro-autoradiography and immuno-fluorescence microscopy. With ultraviolet micro-spectrophotometry changes can sometimes be observed inside the cell, in its nucleic acid or protein regime, and electron microscopy is already directed towards the virus itself. The results of such investigations will be discussed mainly on p. 119.

The anatomical effects of viruses on their plant hosts has been continuously reviewed by Esau (1938, 1948a, 1956, 1961, 1967); she and her colleagues have also greatly contributed to research, especially on ultrastructure (e.g. her book of 1968). For a more recent survey of cytological and histological aberrations in woody plants see Schneider (1973).

Some directly visible chemical changes are in cell pigments, such as chlorophyll, xanthophylls, carotenes, and anthocyanins. In necrosis, black melanins play an important role. These effects will be further discussed in the sections on colour deviations and necrosis.

Another chemical change is *starch accumulation* (see also p. 42), particularly in leaves, by some viruses occurring especially in the phloem of vascular bundles, e.g. potato leafroll, sugar-beet yellows and pea leafroll. Here the leaves usually become thickened and leathery, often brittle, with rolled margins, making the plants rustle when shaken, and completely changing their appearance. Starch accumulation may be restricted to local lesions, visible only after staining (Holmes, 1931), as with chrysanthemum stunt virus in its indicator plant *Senecio cruentus* (Lawson, 1968). Such *starch lesions* may be so small that they are perceptible only under the microscope (Helms & McIntyre, 1962; Fig. 11).

The *excessive formation of callose* in the phloem of stems and tubers of leafroll-diseased potatoes, before necrosis occurs, is of much practical interest. Whereas the sieve plates in phloem vessels of healthy plants are only covered with a thin layer or small plug of callose, the vessels of diseased plants contain large amounts or the cells may even be completely filled (von Brehmer & Rochlin, 1931). Such callose can be stained with resorcin blue or some other agent in stems and even in potato tubers at a certain physiological stage after harvesting (Igel Lange test, see Schuster, 1956). The possibility of using abnormal callose production for diagnosis has been reported almost simultaneously by Igel and Lange (unpublished, but patented in Germany), Baerecke (1955), Hofferbert & zu Putlitz (1955), Moericke (1955) and Sprau (1955); for a survey see Arenz & Hunnius (1963). Schneider (1954) also mentions precocious and excessive callose deposition in the early stages of phloem degeneration in sweet orange with tristeza (quick decline), and Batjer & Schneider (1960) in pear affected with pear decline.

A peculiar chemical deviation directly resulting in a structural abnormality is *defective lignification* of xylem and tracheids of apple, especially in 'Lord Lambourne', after infection with the rubbery wood virus. Cross-sections, treated with phloroglucinol and hydrochloric acid, of diseased branches do not stain (Fig. 13D) or show pale 'islands' in which the secondary layers of the cell walls contain cellulose instead of lignin (Beakbane & Thompson, 1945). The abnormal wood was found to yield more lignin, though the total lignin content was lower than of normal wood, suggesting less firm binding to the polysaccharide frame work of the cell wall (Scurfield & Bland, 1963). This makes the branches very flexible, 'rubbery' and 'cheesy', so that the abnormality has been described as *rubbery wood symptom* (Prentice, 1950b). Older trees develop a 'weeping' appearance as the branches bend under their own weight (Fig. 13B).

Another chemical change is *gummosis*, the production of yellow, orange, or reddish-brown gum-like substances, often preceding or accompanying necrosis and degeneration of various tissues. The gum is generally considered to be a decomposition product of carbohydrates, especially starch, moving into the tracheal elements. It has often been described as wound gum, showing lignin reaction (bibliography in Esau, 1948a). Such gummous degeneration has been observed in psorosis (Fawcett & Bitancourt, 1943) and some other virus diseases of citrus and in attacks of this species by *Phytophthora* spp. and some other fungi or in physical and chemical damage (Klotz & Calavan, 1969). In extreme cases it may lead to the secretion of large quantities of gum (Fig. 44).

More typically structural is *excessive tylosis*, the formation of outgrowths or tyloses from adjacent parenchyma into wood vessels, entering through the pits and swelling into thin-walled bladders in the lumen of the vessel and almost blocking it. Tyloses commonly occur in xylem vessels of grapevine that are no longer functional. They may prematurely occur in summer-collected material affected by leafroll (Hoefert & Gifford, 1967b), and are prevalent in grapevines with Pierce's disease where the blockage leads to growth reduction and wilting especially during rapid growth (Esau, 1948b). This disease is now thought to be caused by a *Rickettsia*-like microorganism.

Characteristic for grapevine leafroll is the *formation of trabeculae*, cellulose bars or rods traversing the lumen of interfascicular and phloem parenchyma cells (Hoefert & Gifford, 1967a;). Such structures have been associated with fanleaf virus infection in grapevine where they most frequently appear in vessels of the xylem (Fig. 14, Gifford *et al.*, 1956) and are also found in tobacco infected with certain strains of tobacco mosaic virus (Wehrmeyer, 1960).

Viruses can have an important effect on size and number of cells, often leading to malformation of organs. An abnormal increase in number is called *hyperplasia* (Gr. πλάσις, plasis = moulding, shaping; from πλάσσειν, plassein = to mould; cf. plasma = something moulded or shaped); when it is almost unlimited, as in tumorous growths, the term (cell) *proliferation* is sometimes used (p. 91). An abnormal enlargement of both cells and organs is referred to as *hypertrophy* (Gr. ὑπέρ, hyper = over, above; τροφός, trophos = feeder; from trephein = to nourish).

Fig. 13 Rubbery wood disease of 'Lord Lambourne' apple (B) due to defective lignification of the wood (D) as indicated by lack of reddening after staining a branch cross-section with phloroglucinol and hydrochloric acid. (From Plant Protection Service, Wageningen.)

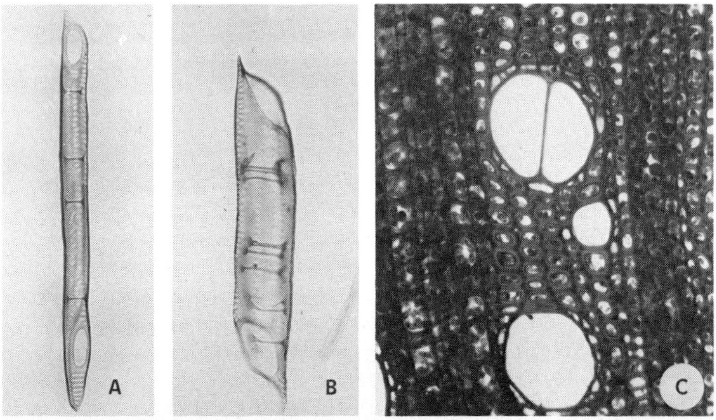

Fig. 14 Trabeculae in tracheary elements of wood of fanleaf-diseased grapevine: (A and B) macerated elements, (C) transverse section of wood (A and B × 110; C × 180). (After Gifford *et al.*, 1956).

The term *hypoplasia* (Gr. ὑπό, hypo = under, less than ordinary) covers development with fewer or smaller cells and underdevelopment of organs, whereas *atrophy* indicates completely arrested growth of cells or organs, a complete lack of multiplication and enlargement where these should occur in the normal growth pattern of the plant.

Certain viruses even influence differentiation of cells in growing points, causing *correlation disturbances* which lead to malformations (p. 87).

A structural deviation, associated with a change in function of certain cells, is *abnormal cork formation* (p. 85).

Loss of turgidity and total collapse of cells are quite common in virus diseases. They are caused by *water deficiency* (p. 73) and *necrosis* (p. 75) or death of cells (from νεκρουν, nekroun = to make dead, to mortify).

It should again be stressed that in certain syndromes various abnormalities are associated. Thus, in yellow leaf areas of plants affected by mosaic viruses, the mesophyll is hypoplastic as compared with normal green tissue, the cells are arranged compactly and there are fewer chloroplasts, which are underdeveloped or even disorganized, and deficient in chlorophyll. Frequently such disorganization kills the cells.

A more complicated and progressive *tissue degeneration* is illustrated by *phloem degeneration.* It has been observed especially with viruses occurring mainly in the phloem and spread by phloem-feeding vectors. Best known are diseases like beet curly top, beet yellows and barley yellow dwarf; though the virus of beet yellows is less strictly limited to the phloem (literature reviewed by Esau, 1967). Strangest of all, the changes brought about by the phloem-bound *Mycoplasma*-like incitant of aster yellows, long thought to be a virus, do not differ essentially from phloem degeneration induced by a true virus (see also Rasa & Esau, 1961). Both with curly top and aster yellows infections the sequence of events is: (1) chromophily of cells bordering sieve elements, with or without accompanying hypertrophy, (2) necrosis of the chromophilic cells, (3) hyperplasia of cells not undergoing primary necrosis, (4) necrosis of hyperplastic tissue, and (5) hypertrophy and hyperplasia of cells surrounding the necrosed tissue.

In both diseases sieve elements deviating in form, size and arrangement predominate. In curly top the abnormalities may even spread from the phloem into cortex and xylem. With barley yellow dwarf, tissue does not proliferate, but the sieve elements and their companion cells, and frequently other parenchyma cells, die, collapse and almost disappear ('necrotic obliteration'). In diseases like potato leafroll, phloem degeneration may be associated with excessive and precocious callose accumulations on the sieve plates. Degeneration of phloem below the bud union in citrus tristeza and pear decline have already been mentioned as the cause of a further progressive debilitation (decline) of the entire tree.

With the help of ultra-microtomes, special embedding and staining techniques, and electron microscopes of high resolution, information has accumulated rapidly on the ultrastructure of virus-infected plants. The most prominent change is chloroplast degeneration which has been studied especially for turnip yellow mosaic virus in Chinese cabbage. Via clumping and fragmentation of chloroplasts the degeneration may vary in severity to complete disintegration. In such affected chloroplasts stroma lamellae are widely scattered and granal stacks are fewer and smaller than normal. Especially striking is the formation of peripheral vesicles, which seem to be invaginations, now thought to play a role in RNA synthesis of the virus (Ushiyama & Matthews, 1970). Other organelles may also be 'visibly' affected in structure and number. Mitochondria may be modified and aggregated as with tobacco rattle virus (Harrison *et al.*, 1970), or enormously enlarged as in *Chenopodium* with an apple virus (Weintraub & Ragetli, 1971). Several of the abnormalities reported are not specific to virus infection, as they also occur in apparently spontaneously degenerating cells or in response to conditions altering the physiology of cells (Shalla, 1968a). Specific, however, are the effects on the nucleus and especially on the nucleolus, which may be very striking, and the inclusion bodies, which are formed by the virus alone or jointly with cell components, or by virus-coded proteins. These abnormalities form the most important subject of submicroscopic research of virus-diseased plants and will be discussed separately (p. 122).

Growth reduction

Many viruses cause a growth reduction similar in proportion in all organs, so that the plants remain morphologically almost normal (Fig. 15). If so, the terms *dwarfing* and *stunting* are applied, as in chrysanthemum stunt (Brierley & Smith, 1951) and in barley yellow dwarf (Oswald & Houston, 1953). The term is sometimes used incorrectly as in *Rubus* stunt, for a witches' broom disease of raspberry (Prentice, 1950a) where it indicates a completely different syndrome (p. 106).

Growth reduction is not always evenly distributed over the plant. This is obvious when the infection has entered in a later stage of development of the host and only the tips of the branches or the top of the plant are stunted, as in pea stunt (Hagedorn & Walker, 1949) caused by red clover vein mosaic virus. A reduction may even be more restricted, e.g. to a part of a leaf with an unevenly distributed discoloration (Fig. 29). This again leads to a malformation.

Growth reduction often gives rise to small fruits, very prominent in little cherry of sweet cherries where it does not become visible before the ripening process starts and results in fruits only half the normal size (Foster *et al.*, 1951). It is the only symptom of this disease which may be due to mycoplasma infection.

Extreme reduction in leaf size is very common, e.g. in various witches' broom diseases such as that of *Vaccinium myrtillus* (Fig. 39; Bos, 1960) and in little leaf of brinjal, *Solanum melongena* (Thomas & Krishnaswami, 1939) (see further p. 106).

The few reports on reduced root growth indicate that it is usually associated with growth reduction throughout the plant.

From the production standpoint, growth reduction resulting in *yield reduction* is most important. It may even be caused by a latent or hardly noticeable virus infection, as with alfalfa mosaic virus and bean yellow mosaic virus in 'Ladino' white clover causing yield reductions of 23 to 55% (Kreitlow *et al.*, 1957).

Tardy development does not always cause low fruit yields. Certain slow-growing rootstocks of fruit trees even improve them, which may explain Posnette & Cropley's observation (1961) of an increase of over 50% in the sweet cherry variety 'Merton Heart' by rusty mottle

Fig. 15 Growth reduction in French bean *(Phaseolus vulgaris)* with bean yellow mosaic virus (left).

and over 100% by rugose mosaic and non-recurrent mosaic. This may be a special case of emergency fruiting, often shown by plants under poor conditions. The reverse may be observed in many plants growing under very favourable circumstances: a decrease in flowering. This means that ecological optima for vegetative development and for flowering are not the same. The example shows that size and yield depend on many factors. Compatibility between scion and rootstock is a similarly complex phenomenon and viruses are known to reduce the success of budding and grafting. One way may be their influence on vegetative vigour of either scion or stock, disturbing the required physiological balance between both (for further details see under incompatibility p. 115).

Colour deviations

Changes in colour, very common in virus-diseased plants, are the first symptoms that have been associated with virus infection, though they are not always the most conspicuous. Most of them are in the chlorophyll-containing outer layers of the plant and therefore the disorders are essentially the same for leaves, young stems and fruits. Since, as a rule, most of the plant's surface is foliage, deviations in leaf colour attract special attention.

Most colour changes are chlorophyll disorders, so that the plants look pale; in only a few cases they are darker than normal, as in oat blue dwarf, characterized by a dark-bluish-green colour (Banttari & Moore, 1962).

The pale-green and yellowish colours indicated by the term *chlorosis* (Gr. χλωρός chloros = light green) are due to a delayed or decreased chlorophyll production. That is why mosaic symptoms may tend to disappear after some time. A further step is degeneration of chloroplasts, as in full-grown leaves of beet infected with beet mosaic and beet yellows viruses (Esau, 1944, 1968). Usually chlorosis accompanies other anatomical disorders, such as spherical palisade cells and delayed formation of intercellular spaces in the mesophyll, both leading to thinner and partly discoloured leaves often further reducing the green colour. According to the surveys by Esau (1938, 1948a), the picture differs according to virus and host. Extreme cases where all the chlorophyll disappears are sometimes called *blanching* or *bleaching*.

In chlorotic tissues the lack of chlorophyll may accentuate the colour of carotenes and xanthophylls, causing *yellowing*. Sometimes the carotene and xanthophyll contents even increase, as in 'White Burley' tobacco plants with tobacco mosaic virus where Venekamp (1957) has found a decrease in chlorophyll from 46.5 to 17.1% together with an increase in carotenes from 35.6 to 45.0% and in xanthophyll from 17.8 to 37.9% in the total quantity of pigments.

It should be kept in mind that the terms chlorosis and yellowing, though as a rule referring to associated phenomena and therefore sometimes equivalent, indicate different processes.

In abnormal *reddening* or sometimes *purpling* of green parts of plants anthocyanins are involved. They also play a part in colour deviations of floral parts, that do not contain chlorophyll. Since anthocyanins are allied to sugars, disturbances in sugar metabolism may be involved. This is supported by the red and purple discolorations often associated with diseases, such as potato leafroll and barley yellow dwarf in oats, caused by viruses limited to phloem disturbing carbohydrate metabolism.

In necrotic tissue the production of dark melanin-like substances results in *browning* and *blackening*. Where necrosis is only superficial, as in the epidermis, the effect may be *bronzing*. If tissue dies by rapid desiccation, melanin may not be produced and the tissue may turn silvery-gray or whitish. Changes due to necrosis and to desiccation will be dealt with in the sections concerned (p. 73 and 75).

Colour deviations in leaves

Colour changes in leaves are multifarious. Many are characteristic for certain viruses, independent of the host, others reflect specific virus-host relationships or vary greatly according to circumstances. Hence their nomenclature deserves critical examination. Because of the confusing variation in colour deviations in leaves, a scheme is given on p. 69 to assist the reader.

The terms *chlorosis* and *yellowing* generally refer to discolorations evenly distributed over the entire plant, the entire leaf, or sometimes part of the leaf or its veins. When a plant is said to have chlorosis or yellowing, this means that the entire plant, or especially its younger parts, have a general chlorosis or general yellowing, as in some deficiency diseases like iron chlorosis and lime-induced chlorosis. This type of leaf disorder is a typical symptom of such well known diseases as beet yellows and barley yellow dwarf and the diseases caused by the very 'polyphagous' beet western yellows virus. They are part of a group of economically most important yellowing diseases ('the yellow plague') which have long been ascribed to various non-infectious factors including mineral deficiencies (Duffus, 1977). The viruses concerned are now classified as luteoviruses (L. luteus = yellow). Mostly, the yellowing appears predominantly in leaves developing after infection. It may then be restricted to the tops of plants or of branches, e.g. top yellows of peas (Fig. 16) caused by bean leafroll virus. It is often expressed as a premature senescence and then first shows up in the older leaves, as in those of head lettuce with beet western yellows virus (Plate 3).

Chlorosis and yellowing often start as an *interveinal chlorosis*, and at a more advanced stage the still green venation may strikingly contrast with the discoloured interveinal tissue as in lettuce with beet western yellows virus.

Chlorosis and yellowing may remain restricted to certain parts of the leaves, suggesting an 'arrested' or incomplete distribution. Examples are chlorosis in leaf tips or barley plants infected with barley yellow dwarf virus and *edge chlorosis* or *edge yellowing* in strawberry leaves, caused by the strawberry yellow edge virus. Both discolorations of leaf tips and leaf edges tend towards a more general discoloration. Another type of incomplete discoloration is that restricted to the veins and known as *vein chlorosis* (Fig. 17) and *vein yellowing*. This discoloration may spread somewhat into the tissue adjoining the veins, but the pattern is always very regular. Good examples are vein yellowing of lucerne caused by bean leafroll virus (van der Want & Bos, 1959; Fig. 18) and yellow net disease of sugar-beet (Sylvester, 1948).

There seems to be a close developmental relationship between general chlorosis or general yellowing and the incomplete types. Such diseases as beet yellows, vein yellowing of lucerne and sugar-beet yellow net are all caused by 'persistent' insect-borne viruses. They occur especially in phloem, causing the vascular tissue, particularly phloem, to degenerate (Esau, 1956, 1957). The fact that bean leafroll virus induces a general chlorosis and yellowing in pea and only vein yellowing in lucerne further proves the relationships between such diseases.

Related to vein chlorosis and vein yellowing is *vein clearing*, where the veins become translucent (Esau, 1933, 1948a) rather than chlorotic or yellow. The process is probably due to a delay in the formation of intercellular spaces caused by cell enlargement and increased cell division, as in curly-top beet leaves, where it may be accompanied by chloroplast degeneration (Esau, 1933). This often makes it hard to

Fig. 16 Chlorosis and yellowing in top of pea plant *(Pisum sativum)* with bean leafroll virus. (After Hubbeling, 1954.)

Fig. 17 Vein chlorosis in pea leaves as a primary symptom of infection with bean yellow mosaic virus (right).

Fig. 18 Bright vein yellowing in 'Du Puits' lucerne with bean leafroll virus. (After van der Want & Bos, 1959.)

distinguish vein chlorosis and vein clearing.

In many diseases vein chlorosis and vein clearing are first or early symptoms, often only temporary.

In contrast to the more regular changes just mentioned, discoloration may also be irregularly distributed over the whole lamina causing *variegation*. Several types of this common phenomenon can be distinguished. Their differentiation may help in reducing the existing confusion, both in describing them and in naming the virus diseases. The classification can be based on differences in shape, and sometimes in size, of the discoloured areas, in sharpness of the boundary between dark and light parts of the leaf, and in distribution of the discoloration over the lamina. As to the colours themselves, the variegation may be chlorotic or even yellow and white, and sometimes even a range in brightness can be indicated. Usually the types of variegation refer to combinations of such differences. The most important types are mosaic, mottling and line pattern.

The term *mosaic* covers a group of symptoms occurring in leaves with a mixture of normal and discoloured areas, usually sharply bordered at small veins. Sometimes the resulting pattern is irregular; in other cases the discoloured areas occur only along veins, or are spread in some other way. It may be fine or coarse, and the abnormal patches may vary from pale-green to yellow or chlorotic.

Typical examples of a well defined mosaic, with a regular distribution of light and dark areas over the whole leaf, are the well known *Abutilon* mosaic (Fig. 2) and tobacco mosaic. The striking discoloration has led to descriptive virus names such as bean yellow mosaic virus, turnip yellow mosaic virus and the more recently described South American whitefly-transmitted bean golden mosaic virus (Costa, 1965).

In 1866, Mayer has introduced the term mosaic for a tobacco disease now called tobacco mosaic which he had proved to be infectious. Since then, mosaic symptoms were found so widespread that for a considerable period virus disease and mosaic disease were considered synonyms.

Mosaic symptoms always start after systemic infection of leaves still beginning to develop. For some reason, certain cells are infected whereas others remain free. Unaffected cells produce a healthy progeny of cells; affected cells an infected, discoloured progeny as illustrated in the experiments of Reid & Matthews (1966) with turnip yellow mosaic virus in chinese cabbage, and by Atkinson & Matthews (1967) with tobacco mosaic virus in tobacco. In both cases the patterns are remarkably constant from the earliest observable stage of development of the leaves till senescence, with the darkgreen 'islands' containing no appreciable amounts of virus.

Sometimes the discoloured areas have a special shape. In monocotyledons, with their parallel-veined leaves, the light-coloured parts tend to be elongate, resulting in streak mosaic or stripe mosaic, as in wheat streak mosaic and barley stripe mosaic. The symptom may be called *streaking* or *striping*, but these terms do not indicate whether it is chlorotic or yellow (as in cocksfoot streak) or necrotic (as in pea streak). Hence, the name leek yellow stripe (Plate 5) is more descriptive.

Other types of mosaic are confined to definite areas of the leaf. In *vein mosaic* the light parts are grouped along the main veins, as in red clover vein mosaic (Hagedorn *et al.*, 1959) and in apple mosaic (Fig. 19A, B). This phenomenon may resemble vein chlorosis or vein yellowing, but the light areas are irregular in shape and distribution, but are always associated with veins, and may include irregularly shaped adjoining tissue. Evidently the association with venation has something to do with the introduction of virus via the vascular tissue at systemic infection.

The term *vein banding* is used for those types of mosaic in which a more regular range of light or sometimes dark tissue occurs along the main veins, as in raspberry vein banding and rose mosaic, but sometimes also with apple mosaic (Fig. 19E). Dark green vein banding often is very conspicuous in cauliflower with cauliflower mosaic virus.

If the mosaic patterns occur mainly between the bigger veins, the name *interveinal mosaic* is often applied, e.g. for symptoms induced by some strains of potato virus X.

Misuse of terms has led to some erroneous names. The name *Abutilon* infectious chlorosis often applied is confusing since it suggests a general pallor of the green; the name *Abutilon* mosaic is preferable and the mosaic may be chlorotic or yellow.

In the literature differences of opinion exist on terms like 'aucuba mosaic' and 'calico' for yellow or very bright mosaics.

The term 'aucuba mosaic' refers to the variegated leaves of *Aucuba japonica* var. *variegata* and is sometimes used to indicate virus diseases, such as potato aucuba mosaic. Since the variegation in *Aucuba* is not always a mosaic but may also consist of flecks or spots, the indication 'mosaic' is inadequate. It also suggests a disease being caused by a mosaic virus from *Aucuba* (Uschdraweit, 1967). The term yellow flecking adequately covers the phenomenon.

The term 'calico' is sometimes used for mosaics with dominant brilliant yellow or even almost white parts of the leaves. Examples are potato calico caused by the alfalfa mosaic virus (Black & Price, 1940), and peach calico (Blodgett, 1944). The name 'calico' is confusing, since in America it is the name of brightly printed cotton cloth, having various colours, whereas in England if refers to plain white cotton cloth. Literally it applies to cloth imported from Calicut (or Kozhikode) in India. Therefore the term 'calico' should be dropped as an indication of a symptom, though it may remain acceptable as a code-name for a disease.

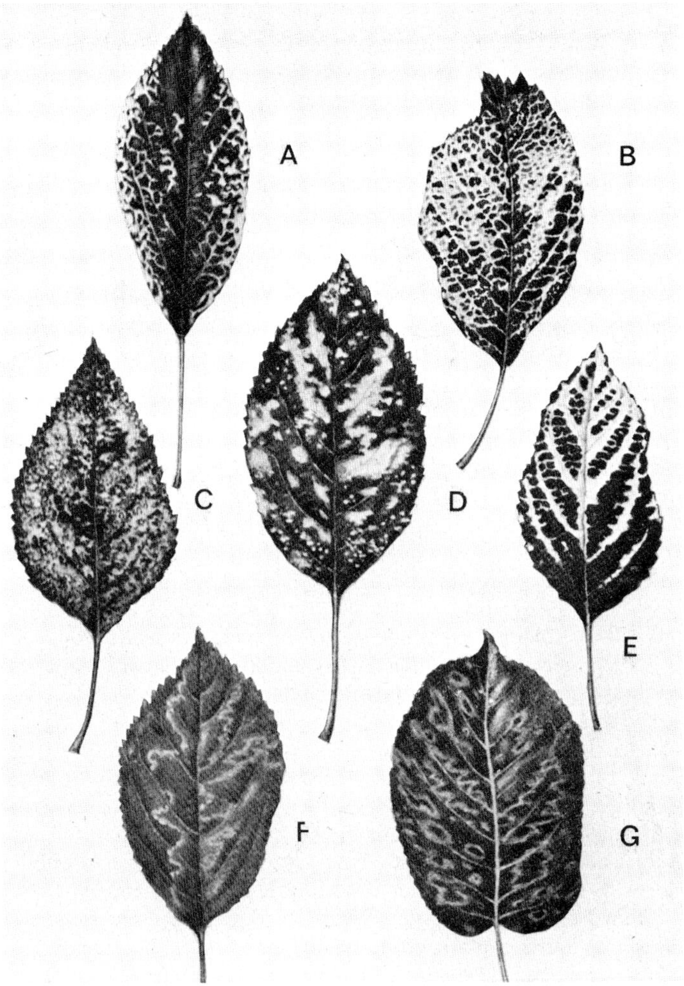

Fig. 19 Some types of variegation ranging from vein mosaic (A, B), yellow mottling (D) and vein banding (E) to line pattern (F, G) with a tendency toward ring formation (G). Leaf G is pear with pear ringspot virus, the others are apple with apple mosaic virus. (From Plant Protection Service, Wageningen.)

If the discoloured patches of a variegated leaf are rounded, the variegation is usually not designated a mosaic but a mottle. In the literature mosaic and mottle are often confused and in some publications they are even considered equivalent. But it seems better to separate them, as many other virologists do, though, as in other forms of variegation, intermediates occur. *Mottling* is very common, e.g. in several fruit tree virus diseases and in some strawberry virus diseases. It is often used to designate virus diseases, such as strawberry mottle (Prentice, 1952).

Various types of mottling, such as *flecking* (Fig. 19D), *dappling*, *freckling*, *speckling*, *specking*, *stippling* (Fig. 20), *spotting*, *dotting* and even *blotching* and *splotching* may be distinguished according to size, shape, distinctness of boundary and number of patches. However many are near synonyms and in normal usage considerable difference of opinion exists about their exact meaning, as is evident from comparison of any two dictionaries. Scientific standardization of these terms seems superfluous as the symptoms themselves may vary considerably with slight differences in external conditions. It is worth noting that the -ling forms are grammatically frequentatives and refer to conditions with numerous spots or patches, whereas the other terms refer either to the spots or patches considered singly, or to conditions with few spots or patches. Certain words, such as blotch, splotch (or splodge) and smudge, just by their terminal sound, refer to large patches with an indistinct boundary. In general these words are best kept for conditions resembling those they commonly describe, e.g. freckles – anything like skin freckles, speckles – anything like on speckled eggs, stippling – anything resembling the effect of dark and shade as produced by an artist with numerous small brush or pencil marks.

Rounded lesions often occur at the sites of inoculation, but also after systemic infection. They may also form light rings around a normally green centre, or even concentric rings of dark and light tissue and this is called *ringspotting* (Fig. 9). Such spots are frequently caused by the important group of ringspot viruses. Discoloration is often accompanied by necrosis. When extending along small veins, the chlorotic or yellow spots may resemble stars or even dendrites (*asteroid spotting*, and possibly *splotching*) which evidently has something to do with the spread of infection along the veins. Light areas sometimes tend to become elongated, as in *streaking*, or linear, as in *striping* of monocotyledons.

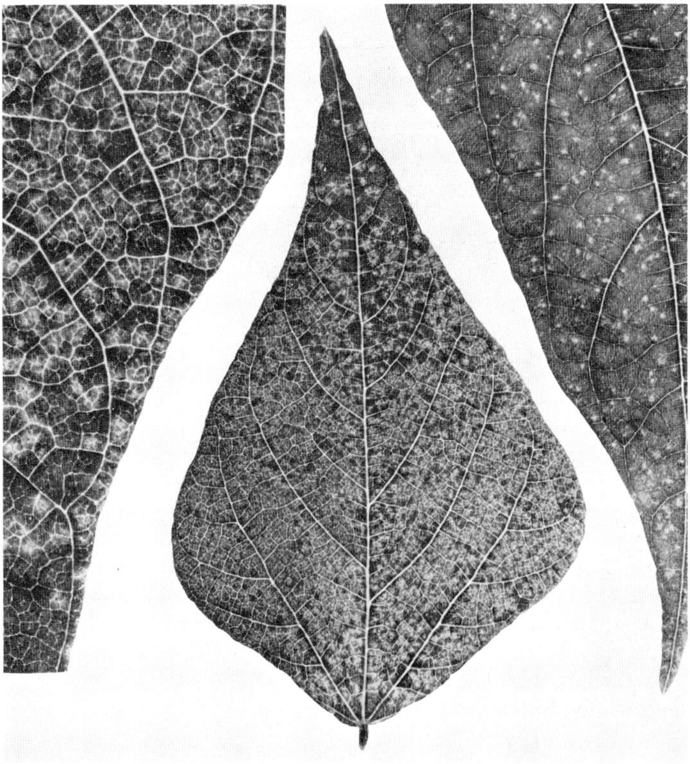

Fig. 20 Yellow systemic stippling in 'Processor' bean *(Phaseolus vulgaris)* with a pea strain of bean yellow mosaic virus. (Left, 2.4 times scale of right and centre.)

Flecks or their surrounding tissue are sometimes somewhat translucent, looking sodden; then they may be named *oil flecks*. They are characteristic of the initial stages of Eckelrader disease (syn. Pfeffinger Krankheit) of sweet cherries (Mulder, 1951) caused by prune dwarf virus and raspberry ringspot virus or *Arabis* mosaic virus.

Sometimes the variegation consists of a conspicuous and often

Fig. 21 Oak-leaf type of line pattern in peach. (From Plant Protection Service, Wageningen.)

brilliant green-yellow pattern formed by irregular single or multiple lines or bands (Fig. 19F, G) indicated as *line patterns*. The lines and bands may also occur as an *oak-leaf pattern* (Fig. 21). A characteristic line pattern is known in various rosaceous plants, such as plum, peach, cherries and roses, diseased by *Prunus* necrotic ringspot virus (Plate 6). The pattern may also consist of irregular rings; unlike ringspotting *ring formation* (Fig. 19G) has normal tissue in the centre of the ring.

In the light parts of the leaf, growth is often impeded, leading to deformation of the leaf (p. 87 and Fig. 29).

A peculiar type of discoloration is involved in gray disease of narcissus caused by the narcissus yellow stripe virus. The silver-gray of the leaves is due to an abnormal number of intercellular spaces (Caldwell & James, 1938).

Entirely different from the chlorophyll disorders described above are *reddening* and *purpling* due to abnormal *anthocyanin formation*, often similar to those induced by mineral deficiencies. They can be observed in many potato varieties after infection with potato leafroll virus (Bode, 1968). Barley yellow dwarf virus causes an intense orange-red colour in oats: oat red leaf (Oswald & Houston, 1953), 'Blattröte' or 'Rotblättrigkeit' (Rademacher & Schwarz, 1958). In carrots, symptoms varying from a reddish tinge in the first leaves to a crimson colour in the whole plant (Plate 4) are due to infection with the persistent aphid-borne carrot red-leaf virus. This virus is usually associated with the non-persistent carrot mottle virus and then their combination is responsible for the carrot motley dwarf syndrome (Watson & Serjeant, 1962). Several other persistent aphid-borne viruses cause such a leaf reddening in few or many hosts such as cotton anthocyanosis virus (Costa, 1956), filaree red-leaf virus (Frazier, 1951) and subterranean clover red-leaf virus (Kellock, 1971), or it is part of their syndrome together with stunting, yellowing and brittleness of leaves. Reddish-purple especially along the edges of the leaves is characteristic of several diseases recently associated with mycoplasma-like organisms, such as potato purple top wilt caused by aster yellows 'virus' (Raymer & Milbrath, 1960).

The production of dark melanin-like substances in dying tissues leads to *browning* and *blackening*. Of course this colour change is very common in virus diseases (see further under necrosis). Pea early browning (Bos & van der Want, 1962) is a typical example of browning. With this soil-borne virus disease veinal necrosis and associated browning spreads irregularly from the vascular tissue in the stem into the leaf veins, causing them to stand out as a brown pattern against the mesophyll of only some leaf parts or leaves. Later on whole leaves and plant tops or the entire plant may turn brown, often early in the growing season. Such restricted brown dendritic leaf patterns occur with a number of viruses upon systemic infection (Fig. 6), though they are most frequent in the inoculated leaf after artificial infection. Browning and blackening are often also associated with the production of ringspots and, of course, with necrotic local lesions (Fig. 3 and 4, Plate 2). If death is caused by rapid desiccation, the production of melanins is prevented, and the tissue may turn silvery-gray or whitish.

Both necrosis and the associated blackening, and desiccation and the associated grayish discoloration play a role in two peculiar phenomena called bronzing and etching.

Bronzing, a rare phenomenon, is due to necrosis and collapse of epidermal cells covering still turgid, green, and apparently healthy mesophyll. It is caused by tomato spotted wilt virus, which provokes circular markings, a network following the finer veins, or almost continuously covers large areas of the tomato leaves. It sometimes leads to general necrosis (Samuel *et al.*, 1930).

Not restricted to the epidermis, but still rather superficial, is the symptom called *etching*. Here shallow necrosis, or collapse of tissue due to local desiccation gives the impression of corrosion, as in tobacco etch disease (Johnson, 1930). The etching often proceeds to a more general necrosis. A number of other viruses may also cause etching (Fig. 22).

The colour changes in leaves are summarized in Table 1.

Fig. 22 Concentric etching in leaf of 'Xanthi' tobacco after systemic infection with cucumber mosaic virus.

Colour deviations in stems

Herbaceous stems can show chlorophyll disorders similar to those in leaves; in peach calico even the twigs show creamy white streaks (Blodgett, 1944). Baker & Campbell (1967) mention a reduction to 40% of the chlorophyll content in bark of young twigs with apple mosaic. With the same disease they report intensified reddening of shoots.

Abnormal anthocyanin formation has attracted attention as a way of diagnosing virus infection in potato tubers. Martin & Quemener (1956) found that 48 hours after transfer of etiolated sprouts 3 or 4 cm long to light, those infected with potato virus X or Y have an irregular

Table 1. Colour deviations in the leaf.

Regular distribution	
general occurrence	intensified green *chlorosis* (weakened green) *bleaching* or *blanching* (disappearance of green) *yellowing* (chlorosis + intensified yellow) *reddening* or *purpling* (anthocyanin formation) *browning* and *blackening* (necrosis) *bronzing* (death of epidermis) *etching* (collapse of superficial tissue by necrosis or desiccation in spots or special patterns)
restricted to certain parts	*top chlorosis* and *top yellowing* *edge chlorosis* and *edge yellowing* *vein chlorosis* and *vein yellowing* *vein clearing* (especially by smaller intercellular spaces)
Irregular and patchy distribution	*variegation*
in sharply defined patches	
patches angular	mosaics: *mosaic*, *vein mosaic*, *vein banding*, *interveinal mosaic*
patches rounded	*mottling: flecking*, *dappling*, *freckling speck(l)ing*, *stippling*, *spotting*, *dotting* *ring formation* (centre normal) *ringspotting* (centre also discoloured but differing from ring)
patches radiating	*asteroid spotting*, *splotching*
patches elongated	*striping* or *streaking*
in diffuse patches	*diffuse mottling*, *blotching*, *splotching* *oil-flecking*
in line or bands	*line pattern* *oak-leaf pattern* *ring formation* *striping*, *streaking*

distribution of pigments, whereas healthy ones are evenly purple or pink. Later, Martin (1958) found that anthocyanin formation in diseased sprouts was strongly retarded during the first three days after transfer to light (i.e. during intensive virus multiplication) but that final content may be higher than normal.

Most colour deviations in stems are due to necrosis (p. 77). Since necrosis frequently originates from the vascular system, the stem is often discoloured by brown or black *streaking*, and even the entire stems may turn black. Herbaceous stems with necrosis only internal, may be discoloured grayish. With bean red node, caused by tobacco streak virus, stem nodes and pulvini of leaves and leaflets turn strikingly red (Thomas & Zaumeyer, 1950), presumably by necrosis.

Because stems have a smaller surface area than leaves, their colour deviations are less prominent than those of leaves, and they are seldom used in disease nomenclature. A rather curious exception is potato stem mottle caused by tobacco rattle virus. Dutch field inspectors have devised this name because, when infected, usually only one or two branches of a plant are abnormal (Rozendaal & van der Want, 1948). Thus the name refers to the irregular distribution of symptoms in the plant and not to the symptoms themselves.

Colour deviations in flowers

Tulips with *flower colour breaking* were first described by Clusius (de l'Écluse) in 1576. They are well known from old Dutch paintings since 1619, and the attractive colours led to a notorious trade in tulip bulbs during the years 1634–1637, known as tulipomania, when fortunes were paid for a single bulb. Of the extensive literature, mainly satyrical, on this subject, the Agriculture University at Wageningen possesses the sole copy of a book by Cos (1637), containing 66 hand-painted drawings with data on bulb weights and prices when publicly sold at Haarlem (Fig. 1). An interesting historical sketch of tulip breaking has been compiled by McKay & Warner in 1933.

Colour breaking, the first virus symptom in plants ever described, is due to local fading, intensification or accumulation of pigments in the epidermal layer of the petals. In fading, the white or yellow colour of the underlying mesophyll becomes visible (Plate 1): *light breaking*. It may give rise to very attractive colour patterns, varying from irregular streaks to fine featherings (see also the colour plates by van Slogteren & de Bruyn Ouboter, 1941). With pigment intensification or *dark breaking*, small dark streaks or elongated flecks develop. The two symptoms often occur together. They cannot be observed in white and yellow varieties since pigments are lacking in the epidermis of their petals. In such varieties, however, infection may be recognized by yellow-red discoloration of the pistillar stigma in fully developed flowers (Elbertsen & Silver, 1973). Both types of breaking have long been ascribed to different viruses, such as a colour-removing and a colour-adding virus (McWhorter, 1938). Now the disease is known to be due to the aphid-transmitted non-persistent tulip breaking virus, sometimes called tulip mosaic virus. The symptoms largely depend on variety and flower development at the moment of infection (Yamaguchi, 1964; Yamaguchi & Hirai, 1967). Some types of light breaking may have a genetic origin. Dark breaking can also be caused by rattle virus (van Slogteren, 1958).

Colour breaking is quite common in the flowers of several other plants, such as gladiolus after infection with bean yellow mosaic virus or cucumber mosaic virus (Klinkowski, 1956) and in stocks (*Matthiola incana*), common wallflower, and several other plant species with turnip mosaic virus (e.g. Tompkins, 1939). Turnip mosaic virus may turn whole petals white. Another well known example is *Iris* mosaic caused by *Iris* mosaic virus (Brierley & McWhorter, 1936). Though some data are available on the effect of virus infection in stocks on the concentration of anthocyanins and other pigments (Feenstra *et al.*, 1963), the biochemical backgrounds of these colour changes are still rather obscure.

Besides these variegative colour changes in flowers, general colour deviations also occur. The colours may be weakened, intensified or entirely changed. After infection with tomato aspermy virus (often called the chrysanthemum strain of cucumber mosaic virus) chrysanthemum flowers with red, bronze or brown colour may turn entirely or partly yellow. Violet-red, light-red or pink flowers may become white-spotted or completely white. Here again, yellow flowers seldom change and white flowers never do (Noordam, 1952).

Greening or *virescence* (L. virescere = to become green) is also a general deviation in colour: chlorophyll causes the petals to be more or less green. Usually this phenomenon is associated with deviations in form and then, virescence is a first stage in the complex of antholysis phenomena (p. 108). An increasing number of such diseases are now associated with mycoplasma-like organisms, so that greening of flowers may not be typical for virus infections.

Colour deviations in fruits and seeds

Colour deviations in fruits are usually chlorophyll disorders like those in leaves and stems. Quite a number are also due to necrotic phenomena, but these will be discussed later (p. 84).

In plants with large fruits, such as gherkin or cucumber, infected with cucumber mosaic virus (Tjallingii, 1952), or cucumber with cucumber green mottle mosaic virus (van Koot & van Dorst, 1959), the colour changes are particularly marked and are often associated with deformations. A pale-green colour associated with reduced growth of cucumber fruits is the most distinctive symptom of infection with the cucumber pale-fruit viroid (van Dorst & Peters, 1974). Tomatoes infected with tomato strains of tobacco mosaic virus may be bronzed, severely mottled, and marbled (Broadbent, 1964).

Also subject to colour changes are many edible pods of legumes, such as French beans infected with various viruses, especially bean southern mosaic virus and bean pod mottle virus. With the latter, pods of many varieties are severely mottled and darker than normal, and usually also deformed (Zaumeyer & Thomas, 1948). With bean common mosaic virus the bean pods may show a pale green slightly sunken stripe along the entire pod in cultivars reacting with a dark malforming banding along the main leaf veins.

Other pigments than chlorophyll may be involved in abnormal fruit colour. With tomato spotted wilt in tomatoes the deviations may be very pronounced: the ripe fruits are usually paler red, often yellow or, more rarely, white patches may occur. The pale patches vary in shape from an irregular mottle to distinct concentric circles, or almost the entire skin may be yellowish with small islands of normal red (Samuel *et al.*, 1930). Red bands and thin red rings and lines on plum fruits, that normally become orange, red or purple during ripening, were found of value to diagnose sharka infection, whereas grooves and pits previously thought to be the main symptom were considered unreliable (van Oosten, 1972).

Colour abnormalities in seeds are often overlooked. They occur, for instance, in soya where the normally uniformly coloured beans may be speckled or mottled after infection with the seed-borne soybean mosaic virus (Koshimizu & Iizuka, 1957; Ross, 1968). A faint spotting and some wrinkling of pea seeds has been reported for pea early-browning virus (Bos & van der Want, 1962). These and some other seed abnormalities are not directly correlated with seed transmission of the viruses concerned. The seed-coat is part of the mother plant and mother-plant infection does not necessarily lead to embryo infection (for a survey see Bos, 1977).

Water deficiency

Diseased plants often show *water deficiency* causing loss of turgidity leading to *wilting* or a total loss of water, called *desiccation*, and *withering* (desiccation associated with shrinking and shrivelling). In some languages there are no terms to separate desiccation and withering. In wilting the tissue shows a flaccidity from which the plant may recover, though usually only temporarily. Wilting often proceeds to withering which is irreversible and is accompanied by death.

Withering is rather common in virus diseases, e.g. in peas with early browning, where it happens irregularly over the plant as a consequence of irregular necrosis of vascular bundles in veins, petioles and stems. With many viruses, withering can be observed in mechanically inoculated leaves.

Desiccation may be confined to certain parts of the leaf. In *Phaseolus vulgaris* infected with tobacco necrosis virus (bean stipple streak) the smallest veins in restricted areas of the lamina may be necrotic causing desiccation of the enclosed laminar tissue. Because the dry area is surrounded by normally turgid leaf tissue no shrivelling occurs. Local lesions often consist of desiccated tissue and then they are gray or whitish; in inoculated leaves their coalescence may result in large dry areas.

A peculiar type of desiccation is *etching*, where local collapse of superficial tissue gives the impression of corrosion. This collapse may be due to desiccation or to necrosis (see also p. 77).

In gherkins or cucumbers wilting due to cucumber mosaic virus is very striking when, during the second week after infection, the weather is cloudy and cool with maximum day temperatures not over 20°C (Tjallingii, 1952; Fig. 23). Wilting at noon has been reported for sugar-cane with sugar-cane chlorotic streak virus (see Holmes, 1964).

Sometimes wilting has been incorporated in the name of a disease or virus. Pea wilt is a severe disease in garden and field peas caused by white clover mosaic virus (Bos *et al.*, 1959). With broad bean vascular wilt, caused by a virus of the same name, plant tops first

Fig. 23 Wilting proceeding from the stem tip downward in gherkins *(Cucumis sativus)* with cucumber mosaic virus. Left, healthy plant. (After Tjallingii, 1952.)

become flaccid and blackened; gradually the whole plant loses turgidity, and may collapse and die (Stubbs, 1947).

As yet little is known about the immediate cause of water deficiency after virus infection. It must involve either reduced supply of water or excessive transpiration. Reduced supply may be due to necrosis of vascular bundles or to impregnation of vessels and other xylem cells with gum. In gherkins with wilting caused by cucumber mosaic virus, where wilting starts in the tops of the stem and proceeds downward, Tjallingii (pers. commun.) could not find any irregularity in the vascular system. This suggests excessive transpiration.

Necrosis

Loss of water and various degenerative changes leading to death of cells and tissues have already repeatedly been mentioned. Such death, not directly killing the whole plant, is called *necrosis*. In the past, when virus disease and mosaic disease were still considered synonymous, necrosis symptoms were usually overlooked, despite their commonness after virus infection.

Necrosis is usually rapid, with a clear borderline between dead and live tissue, especially where dark melanin-like substances occur in the dead cells (see also colour deviations, p. 67). Where such discoloration is absent, death may be accompanied by rapid desiccation, or desiccation may even be the cause of death, preventing chemical changes in the protoplasm that otherwise produce melanin.

Site and type of necrosis often have diagnostic value: it may affect superficial cells; it may occur in a deeper tissue; it may involve several tissues, or it may be restricted to one. Some anatomic data will be given when discussing necrosis in stems.

Necrosis frequently develops at the site of virus entry as a shock reaction, confined to the inoculated cell and some surrounding cells, resulting in a necrotic *local lesion* (Fig. 3, left). Such a shock reaction, demonstrating *hypersensitivity*, often prevents systemic infection (p. 18).

Where necrosis primarily involves essential tissues, as in vascular necrosis, the entire plant may soon succumb (see also decline, p. 25).

Under humid conditions, necrosis often allows *rotting* by secondary fungi or, more often, by bacteria, but this is a phenomenon due to a secondary cause (p. 118).

Necrosis in leaves

Many viruses, in a multitude of plant species, cause *necrotic spotting* or *necrotic speck(l)ing* in the interveinal tissue after mechanical inoculation. As mentioned before, Holmes (1929) first used the number of local lesions of tobacco mosaic virus in *Nicotiana glutinosa* leaves to estimate the virus concentration in the inoculum. Test plants widely used for this purpose are *Chenopodium amaranticolor*, *C. quinoa*, *Gomphrena globosa*, *Nicotiana glutinosa*, *N. tabacum* 'White Burley' and 'Samsun', *Petunia hybrida*, *Phaseolus vulgaris*, *Solanum demissum* 'A6', *Tetragonia tetragonoides* (= *T. expansa*), *Vigna sinensis* and

Zinnia elegans. Several of these react to various viruses.

Necrotic lesions vary in size and number with virus and external conditions; their colours range from brown, red or reddish to black. In *necrotic ringspotting* they consist of concentric rings of necrotic, yellow and desiccated tissues. If the symptom is merely a pattern of rings and arcs with normal tissue in the centre, it can be termed *necrotic ring formation*.

Necrotic spots, rings, and ringspots may also occur after systemic infection when only a few infectious particles succeed in establishing secondary infection centres. A good example of such systemic necrotic spots is necrotic stipple in stored cabbage caused by cauliflower mosaic virus (van Hoof, 1952; Fig. 24). In various plant species local and systemic ringspots are evoked by ringspot viruses like tobacco ringspot virus and beet ringspot virus. Prominent systemic brown and black streaks, rings, and ringspots are also found in orchid leaves with

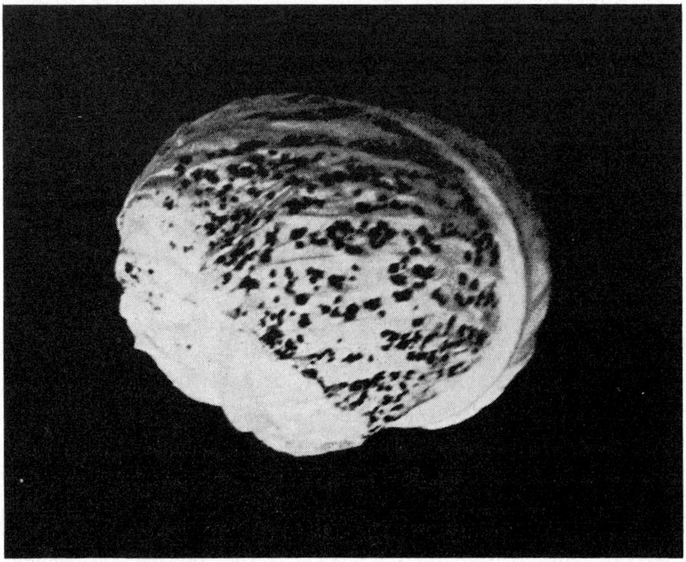

Fig. 24 Necrotic stipple of common cabbage developing during storage due to systemic infection with cauliflower mosaic virus. (From Dr H. A. van Hoof, Wageningen.)

Cymbidium mosaic virus (formerly called *Cymbidium* black streak virus; Jensen, 1951) and *Odontoglossum* ringspot virus.

With necrotic rusty mottle in cherry leaves, the necrotic areas frequently fall out late in the season, causing a conspicuous *shot-hole* effect (Richards & Reeves, 1951).

Where necrosis is only superficial and more diffuse, the effect is that of corrosion, often called *etching* (Fig. 22). But such superficial damage may also be due to shallow desiccation of tissue. The term *bronzing* is used when necrosis and collapse are restricted to epidermal cells overlying still turgid and green mesophyll. Both symptoms have already been mentioned under colour changes of leaves.

When a large number of necrotic lesions develop after mechanical inoculation, they may coalesce and form dead areas. Sometimes the necrotic lesions gradually or rapidly enlarge, producing a more systemic necrosis, as in some streak diseases of peas, where the distribution of necrosis is irregular. But ultimately the whole leaf may die. After reaching the veins, and spreading rather quickly within the vascular system of the leaves, these viruses often start with *veinal necrosis*, after some time proceeding to the interveinal tissue, and desiccation of this tissue may result, as in pea plants with early browning. A spreading necrosis of the veins is also caused by tobacco necrosis virus in beans, mainly in inoculated leaves (bean stipple streak), and with some other viruses (Fig. 6).

Necrosis in stems

After reaching the veins, necrosis generally does not remain confined to the leaf but spreads along the petioles to the vascular system of the stem from there to the higher leaves. It then often disturbs water supply and causes wilting and withering. The way necrosis is distributed in the vascular system is frequently considered illustrative of the translocation of virus in the plant. Such a necrosis can be induced in a hypersensitive host as *Nicotiana tabacum* 'Xanthi-nc' by introducing the virus (tobacco mosaic virus) into the vascular tissue by grafting from *N. tabacum* 'Samsun'. It can also be evoked by transferring inoculated plants to 24 °C after systemic infection at 30 °C (de Leeuw, 1968).

Such systemic vascular necrosis is the main symptom in black root (Jenkins, 1941) in some varieties of snap bean, field resistant to bean common mosaic virus. In these varieties the virus normally remains confined to small necrotic spots, but at temperatures above 20 °C it may become systemic and induce necrosis in vascular tissues of all plant parts and become especially conspicuous in roots, stems and pods. The same systemic vascular necrosis occurs in these hypersensitive varieties at normal temperatures if the virus is introduced into the vascular system by grafting (Grogan & Walker, 1948). Later strains of the virus were isolated that easily induce such a local and systemic necrotic vascular reaction at 20 °C in the varieties with dominant resistance derived from 'Corbett Refugee' (Hubbeling, 1972) (Plate 7).

Quite commonly systemic reaction leads to death of young sprouts or tops of stems, as in beans after infection with a special strain of bean yellow mosaic virus, and in some potato varieties after infection with potato virus A or X. This *top necrosis* (Fig. 25) is sometimes called acro necrosis. Occasionally tomato bushy stunt virus causes it in tomato plants, where death of the growing points is followed by development of secondary shoots, which give the plant a bushy appearance (Smith, 1935; see also Fig. 7).

Fig. 25 Severe top necrosis in 'Ambassadeur' potato with potato virus X. (From Ir A. Rozendaal, Wageningen.)

In yellow lupin infected with cucumber mosaic virus, necrosis turns stems and especially tops brown; they become distorted and brittle and easily break. Such apical death is common in *Chenopodium amaranticolor* and *C. quinoa* inoculated with a number of viruses including beet mosaic virus. Sometimes the effect starts on one side of the stem, causing their tops to curve sideways. A comparable syndrome occurs in soybean bud blight due to tobacco ringspot virus (Allington, 1946).

Necrosis in stems may originate in various ways.

In diseases caused by viruses occurring mainly in the phloem, necrosis is, as a rule, restricted to this tissue. A classic example is *phloem necrosis* in potato plants infected by leafroll virus (Quanjer, 1913). That necrosis is visible only under the microscope and involves sieve tubes and companion cells. The phloem necrosis sometimes proceeds to the tubers (Fig. 27A).

In pea top yellows, caused by bean leafroll virus, cross-sections of the stem show phloem necrosis (de Fluiter & Hubbeling, 1955), readily visible macroscopically after superficial scraping off the cortex as a reddish colour at the stem basis by which it can be distinguished from attacks by *Fusarium oxysporum* f. sp. *pisi* causing a reddening of the more internal xylem. Two other examples of phloem necrosis are sugar-beet with curly top (Esau, 1933), and grasses (including cereals) infected by barley yellow dwarf virus (Esau, 1957). Phloem necrosis is of special importance in such diseases as citrus tristeza and pear decline, where it occurs immediately below the bud union; with these diseases the effects accumulate, finally resulting in an externally visible decline syndrome throughout the plant as described under sequence of symptoms (p. 25).

Pear decline is now known to be due to mycoplasmas and their effects on vascular tissues may closely resemble those of phloem-limited viruses (see p. 53). Another mycoplasma-incited disease, elm phloem necrosis (Swingle, 1938), even derived its name from the symptom.

Because the virus is transported through the phloem, many types of necrosis start there; later they may extend to other tissues. In the black-root disease of beans already mentioned, necrosis also affects the cambium and the outer layer of the xylem (Jenkins, 1941). With several viruses occurring solely in the phloem, vascular necrosis is associated with a complex of progressive degenerative changes, as mentioned under anatomical deviations (p. 53).

In the streak diseases of potatoes, grouped under the name 'acro necroses' by Quanjer (1931), necrosis arises in the phloem and then spreads in all directions into the neighbouring tissues, most markedly towards xylem. Internal necrosis is often visible to the naked eye as a diffuse dark *streaking* on stems, petioles and main veins.

Stem necrosis may also find its origin in the parenchyma. In stems, petioles and main veins of French beans infected with white clover mosaic virus, parenchyma cells in the pericambium (the tissue between phloem and cortex), or between xylem elements or groups of interfascicular parenchyma cells may be necrotic (Bos, unpublished data). This necrosis, preceded or accompanied by gum deposition, is externally visible as dark-grayish streaky discolorations.

In potato plants affected by potato virus Y (leaf drop streak, stipple streak or acropetal necrosis), necrosis occurs in the collenchyma of the aerial organs and sometimes extends to other tissues of the cortex, but not to vascular bundles. In petioles even the parenchyma between bundles may become affected (Quanjer, 1931). The necrotic streaks are visible from the outside. In tobacco stems infected with tobacco rattle virus, the pith and the cortex show considerable necrosis (Böning, 1931).

Besides such internal necrosis, a stem can show more superficial necrotic symptoms, almost restricted to the cortex, resembling necrotic lesions in leaves and petioles. For example, in potato stem mottle, caused by tobacco rattle virus, it starts in the leaf and proceeds to the cortex of veins, petioles and stems without affecting vascular bundles (Quanjer, 1931).

In the literature the term 'streak' is often used for these necrotic areas as well as for the diseases concerned. Without further specification this is rather confusing, because there are several 'streak diseases', as cocksfoot streak, characterized only by elongate chlorotic discolorations (see also under colour deviations, p. 61). Several necrotic streak diseases have been described, e.g. for peas, and in this crop some eight different viruses have been found as incitants.

In woody stems the rather common *bark necrosis* is frequently associated with *canker*. According to a literature review by Zycha (1955) canker has never been precisely defined, hence some confusion. These and some others terms should be examined.

With elm zonate canker bark necrosis appears as alternating dead and live zones starting in the cortical or phloem tissue. Later the dead rings enlarge and necrosis may extend into the xylem and cause the bark to burst, stems and branches may be girdled and their upper parts may die (Swingle & Bretz, 1950). Similar disorders occur in some cherry varieties infected with sweet cherry necrotic mottle virus (Richards & Reeves, 1951), where they may develop into numerous cankerous or blister-like lesions in the cortex of young branches and later cause pronounced roughening of the bark. Similar symptoms are those of pear blister canker (Cropley, 1960; Plese *et al.*, 1971; Fig. 26) and prune diamond canker (Smith & Thomas, 1951), the latter accompanied by swelling.

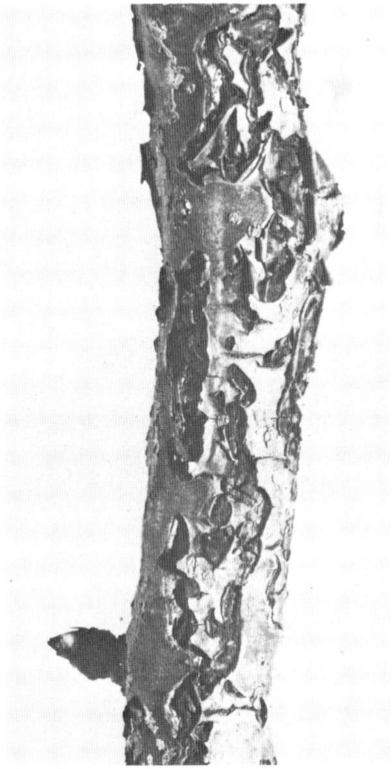

Fig. 26 Pear blister canker on twig of 'Williams Bon Chrétien' pear. (After Cropley, 1960.)

In most of these diseases, only the bark is necrotic, or this phenomenon is accompanied by swelling and ulcerous blackening. Because the abnormal tissues are likely to decay, plant pathologists tend to apply the term canker. The decay is due to secondary rotting of dead tissue and this contributes greatly to the black ulcerous and crusty appearance of the wound. Some fungal diseases with such blackening are sometimes called anthracnose (from Gr. ἄνθραξ, anthrax = coal, νόσος, nosos = disease). In English, a distinction is often made between 'annual' and 'perennial' cankers: the first do not spread and should be called bark necrosis, the latter are chronic and continue to enlarge.

Necrosis may also ensue from a bursting of stem tissue, mechanically caused by local swelling, as in 'Napoleon' sweet cherry with cherry black canker (Fig. 35; p. 97).

Obviously, allied abnormalities range from mere bark necrosis with dead and dry black tissue, through various types of canker where surrounding tissues react and later may also become involved, through decay of dead ulcerous tissue, to tumours where only new (but disorganized) tissue is formed (see further p. 95). It seems appropriate to distinguish within this range between *tumorous cankers* for necrotic lesions in bark of woody plants where swellings are clearly secondary, and *cankerous tumours* where swelling is primary and necrosis and decay ensue later (p. 98).

Potato tubers may show a series of mainly internal necrotic phenomena considerably influencing consumption quality.

In corky ringspot or 'spraing' (Scots Doric word for a bright streak or stripe), caused by tobacco rattle virus, the transverse section of a tuber shows arcuate or ring-shaped necrotic patterns, a type of *ring formation* (Fig. 27C, D; for literature see Eibner, 1959, and Walkinshaw & Larson, 1959). The name corky ringspot refers to some accompanying cork formation; the untranslatable Dutch 'kringerigheid' points to the irregularity of the rings, the German name 'Pfropfenbildung' (formation of pellets or wads) expresses the three-dimensional nature of the symptom (Fig. 27C). The origin of these spatial structures is similar to that of the two-dimensional necrotic rings in leaves. Recently, the symptom has also been associated with potato mop-top, another soil-borne disease, but there slightly raised necrotic or partially necrotic concentric rings are found on the surface of the tuber (Calvert & Harrison, 1966).

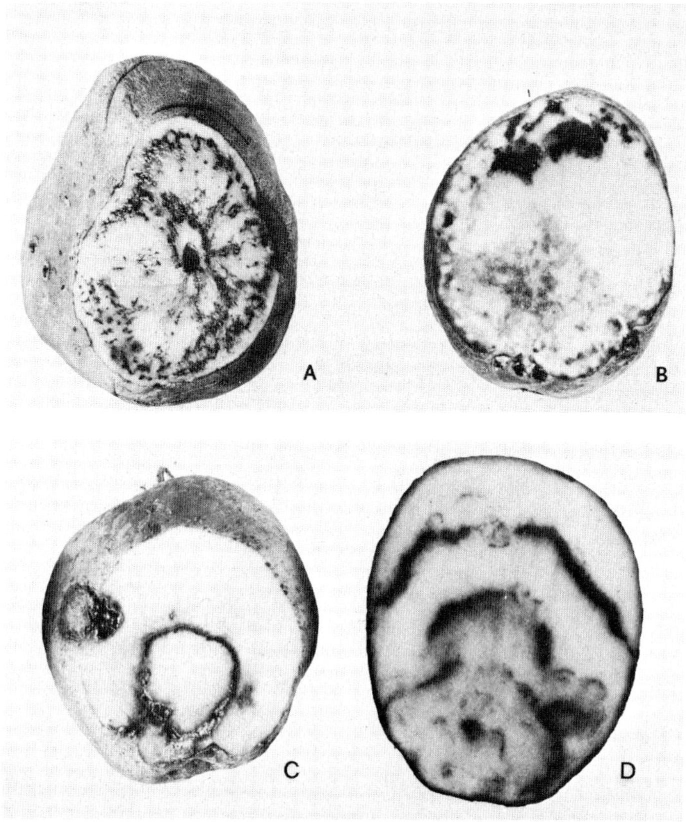

Fig. 27 Some types of necrosis in potato tubers: net necrosis with potato leafroll virus (A); pseudo net necrosis with potato aucuba virus (B); primary symptoms of corky ringspot with tobacco rattle virus (potato stem mottle) after infection from the soil via the skin (C) and secondary symptoms of corky ringspot after infection via the hilum (D). (A. After Folsom et al., 1938; B. and D. from Ir A. Rozendaal, Wageningen; C. from Plant Protection Service, Wageningen.)

In corky ringspot, Lihnell (1958) distinguishes between primary and secondary symptoms, according to their position in the tuber. The primary symptoms, often seem to emanate from a centre at the surface of the tuber (Fig. 27C), the secondary symptoms are often restricted to the heel end and are mostly arranged around the hilum (Fig. 27D). The difference can be explained by assuming that in the first case the incitant has entered from the soil through the skin, in the second case systemically through the stolon.

The tubers of some potato varieties, especially in North America, show 'net necrosis' after infection with leafroll virus. In their subsurface tissue dark-brown flecks, stripes and reticulations develop, consisting of necrotic sieve tubes and companion cells. This *phloem necrosis* in the irregularly distributed vascular bundles is plainly visible to the naked eye (Fig. 27A; Folsom *et al.*, 1938). Tuber *blotching* (sometimes called pseudo net necrosis), usually occurring in the parenchyma of both cortex and pith, is caused by potato aucuba mosaic virus (Fig. 27B). It is readily visible as rusty to dark-brown flecking and stippling within and outside the vascular ring (Clinch *et al.*, 1936).

The phenomena mentioned above differ from 'tuber rust spot', the German 'Eisenfleckigkeit', in the less diffuse spots, which are assumed to be of physiological origin (Eibner, 1959). But Cervantes & Larson (1961), and others, have found such potato tuber disorders to be caused by alfalfa mosaic virus.

Necrosis in fruits and seeds

Necrosis in fruit greatly influences its marketing value because the abnormalities are defects of appearance and influence consumption quality and shelf life.

Cucumbers may be covered with slightly necrotic spots surrounded by a watery dark-green zone followed by premature rotting after infection with cucumber necrosis virus (van Koot & van Dorst, 1959). In green pods of French beans with bean common mosaic virus, suture and pod wall may show severe necrotic discoloration caused by a vascular necrosis also apparent in stems and roots (black root disease). According to Grogan & Walker (1948), it occurs especially in hypersensitive varieties at high temperatures, but Hubbeling (1972) has also observed it at normal temperatures with abnormally virulent strains. With tobacco necrosis virus (bean stipple streak) such green pods are covered with rusty flecks or ring-like patterns also affecting internal tissues (e.g. van der Want, 1948). Similar, but blacker and more superficial lesions on bean pods can be caused by bean yellow mosaic virus.

Quite a number of streak-inciting pea viruses are able to evoke necrosis in pea pods varying from necrotic spotting to purplish coloration of the entire pod, such as cucumber mosaic virus (Whipple & Walker, 1941) and beet mosaic virus (Quantz, 1958a). With pea early-browning virus (Bos & van der Want, 1962) the pods often show irregular brown to purplish flecks and rings, without any effect on leaves and stems. With pea streak virus, Ford (1964) has shown that pea plants infected early can produce symptomless pods because the pods develop when the plant has already entered a chronic stage of infection. Plants infected about four weeks after planting, or later, develop pods with progressively more necrosis and foliage with less necrosis.

Another example of necrotic rings in fruits is tomatoes infected by a necrotic strain of tomato spotted wilt virus (Kovačevski, 1959).

Several tree-fruits have been reported to show necrosis after virus infection. In pears with stony pit virus, necrotic centres occur in the fruit in addition to concentrations of sclerenchyma cells (Kienholz, 1939). Very striking are the large concentric, sometimes coalescent, necrotic rings developing on the skin of apple fruits when infected with apple ringspot virus at the end of growth (Canova, 1963).

Reports on necrosis in seeds are rare, though early necrosis frequently causes *seed abortion.* This subject will be discussed more in detail in the section on miscellaneous abnormalities.

A disease of broad bean *(Vicia faba)* has been described in England as Evesham stain (Lloyd *et al.*, 1966), characterized mainly by local brown necrosis of the testa, chiefly in a ring around the periphery of the seed. The virus has later been named broad bean stain virus.

Abnormal cork formation

Cork formation is quite normal in plants, so here only its abnormal occurrence has to be considered, but it has been rarely studied. It involves no production of abnormal cells but only abnormal numbers of suberized cells, so that it may be called an organizational disturbance.

Klinkenberg (1940) has investigated the anatomy of abnormal

cork in the roots of *Lupinus polyphyllus* affected by sore shin, presumably in that material caused by cucumber mosaic virus. The cork forms a continuous layer at the base of small swellings on the roots and, in deeper layers, around intercellular spaces filled with gum or around groups of necrotic cells.

In psorosis (from Gr. ψώρα, psora = scab) of citrus, development of cork in the bark of stems cuts off the outer layers, causing *bark scaling* (Fawcett & Bitancourt, 1943). With citrus exocortis, scaling is induced by development of successive layers of wound periderm immediately under lesions with necrosis, hypertrophy and hyperplasia in the bark. In trifoliate rootstocks large patches of outer bark may scale off (Schneider, 1973). Abnormal layers of periderm are also involved in pear blister canker (Fig. 26; Plese *et al.*, 1971).

A peculiar and conspicuous type of cork formation is *rough skin* on apples infected with apple rough skin virus, resulting in brown patches on the fruits (Fig. 28 left). These patches may be small and sometimes almost circular (Plate 8), or they may occur as streaks, while on fruits of severely affected trees, the patches cover large areas of the skin. Sometimes these patches are cracked and the fruits are slightly deformed by local growth retardation (van Katwijk, 1955, 1956). *Star cracking* in the corky patches occurs with the, presumably related, apple star cracking virus (Fig. 28 right; Jenkins & Storey, 1955).

Fig. 28 Apple rough skin in 'Glorie van Holland' (left) and apple star cracking in 'Golden Delicious' (right). (From Institute of Phytopathological Research, Wageningen.)

Malformations

Like abnormal cork formation, *malformation* is caused by disturbance of harmonious tissue development, but here it leads to abnormalities in plant form: morphogenesis is disturbed. Malformations differ from stunting and dwarfing which are due to a general growth reduction discussed earlier (p. 54). Together with growth reduction they represent abnormal growth.

Malformations are deformations or changes in shape that are pathological. They are of many types. Often cytological changes are also involved, so that the borderline with other groups of symptoms is not always sharp. They can be grouped in various ways. The way chosen here (Table 2) is to distinguish between primary and secondary malformations. When growth correlation is more directly disturbed by virus infection, the malformations are primary (for primary see p. 22). Secondary malformations become evident only after the plant has developed other symptoms that in turn lead to the malformation. The primary malformations can be ascribed to hormonal disturbance, whereas the direct causes of secondary malformations are more mechanical. Again, the distinction is not always sharp and both may be associated.

Of course, the effects of a virus in a single plant may vary with time. Thus, developmental morphology may be aberrant (Takahashi, 1972). By time-lapse cinematography, Novák (1964) found that spiral growth rhythms were especially influenced in the youngest and middle leaves of tobacco infected with tobacco mosaic.

Secondary malformations

Secondary malformations are, so to say, natural consequences of stresses developing after certain primary symptoms and they illustrate the complexity of disease, where one effect may lead to another. They are rather common in virus diseases.

Discoloured parts of a leaf often grow less than normal areas. An irregular mosaic may cause internal tensions, leading for instance, to abnormally deep incisions or even to irregular lobes of the leaf margin when the mosaic occurs near to it.

Various types of *leaf narrowing* may result. Most of them usually give the impression of a hormonal disturbance and they will therefore be discussed under primary malformations. Often the tensions are

Table 2. Survey of malformations[1].

Secondary malformations (mechanical consequences of other virus symptoms)

- leaf malformations
  - *leaf narrowing* (in part)
  - *crinkling*, *curling* (in part)
  - *distortion*
  - *leafrolling* (in part)
- stem distortion
- bushy growth
- fruit malformations (in part)

Primary malformations (results of hormonal imbalance)

- *proliferation*
- *histoid malformations*
  - localized overdevelopment of certain tissues
    - *histoid enations*
    - *tumours: cankerous tumours, tumorous cankers*
    - *shoot swelling*
  - localized underdevelopment of certain tissues
    - *stem pitting*
    - *stem grooving*
    - *branch flattening*
- *organoid malformations (teratomata)*
  - in leaves
    - *leaf narrowing* (in part)*: shoe stringing, fanleaf formation*
    - *rugosity*
    - *crinkling* and *curling* (in part)
    - *epinasty*
    - *leafrolling* (in part)
    - *hypertrophy of stipules* (due to mycoplasma?)
    - *organoid enations*
  - in stems
    - *internode shortening: double nodes, rosetting, zigzag growth*
  - in entire plants
    - *witches' broom phenomena* (due to mycoplasma?)
      - *witches' broom growth*
      - *antholysis: virescence, phyllody, apostasis, prolification, sterility*

---

[1] Malformations + growth reduction (dwarfing, stunting) = abnormal growth.

released by twisting or curvature out of the leaf's plane, resulting in symptoms as often induced by tobacco mosaic virus in tobacco leaves (Fig. 29) and by bean common mosaic virus in leaves of French beans. In the latter the dark-green areas, mostly along the midrib, grow faster than adjacent tissue into vermicular bulges. Such diseased leaves are often so tortuous that description is impossible with morphological terms.

Local growth reductions in leaves, especially in the veins, often cause irregular leaf deformation, such as *crinkling* and *curling*; internal torsion forces may lead to *distortion*. Such malformations may be due to local necrosis in mesophyll or in veins. Since necrotic tissue no longer grows, tensions are induced in surrounding expanding tissues.

Some persistent aphid-borne viruses occurring mainly in the phloem of their hosts may exert another drastic effect on plant growth by causing accumulation of starch. Besides becoming thickened, leathery and brittle, and rustling when touched, leaves affected with potato leafroll usually also show *leafrolling*, an upward curling of their edges. Barley plants with yellow dwarf are stunted, leaves are erect, thicker and stiffer than normal, and tillering is usually stimulated (Oswald & Houston, 1953).

Like leaves, stems (and especially their tips) may curl or become distorted by local necrosis, as with a number of viruses in artificially

Fig. 29 Secondary leaf malformation after mosaic in 'White Burley' tobacco with tobacco mosaic virus.

infected *Chenopodium* species. With complete necrosis of the stem apex lack of apical dominance may cause development of lateral branches. When these in turn are attacked, branching proceeds, resulting in a bushy plant like a witches' broom as in tomato aspermy (Blencowe & Caldwell, 1949). Similarly, in tomato, necrosis occasionally develops after infection with tomato bushy stunt virus that kills the growing points. Secondary shoots then develop producing the bushy or rosetted appearance of the plant after which the disease is named (Smith, 1935). See also Fig. 7.

A growth reduction like that in foliage leaves may be the cause of various deformations in flowers of *Datura stramonium* with certain strains of cucumber mosaic virus (Juretić, 1968). They consist of shortening, narrowing, irregular development, distortion, abnormal number of floral parts (including stamens and carpels), open pistils, and poor development of spines on fruits.

Related malformations, starting as a mosaic, may occur in fruits, as in gherkins after infection with cucumber mosaic virus (Fig. 30; Tjallingii, 1952) where dark-green tissues that have grown normally protrude from the fruit's surface. Localized necrosis and excessive formation of sclerenchyma in stony pit of pear also leads to mal-

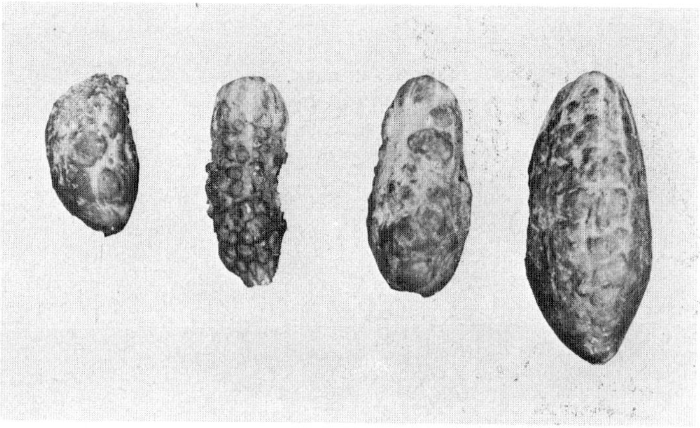

Fig. 30 Secondary malformation in gherkins *(Cucumis sativus)* with cucumber mosaic virus. (After Tjallingii, 1952.)

formation of fruits (Kienholz, 1939). In raspberry, crumbly fruits may result from an irregular failing of drupelets to set as a consequence of seed abortion due to infection by some viruses but sometimes also to genetic factors (Murant *et al.*, 1974).

Primary malformations

As remarked, primary malformations are those resulting directly from disorganization of growth by the virus, or at least more directly than secondary malformations. They may be described as products of abnormal morphogenesis and are probably due to aberrant action of plant hormones (Bos, 1957b, 1970) e.g. by changes in distribution or by changes in general concentration of hormones. Unfortunately, data on both quantitative and qualitative influences of virus infection on growth hormones are still scarce.

Real malformations can be divided into two groups, as Küster (1911, 1925) has done for plant galls incited by animals and fungi. The *histoid* or *histological deviations* (histoid = like tissues) are due to abnormal organization within tissue, or within an organ. In *organoid* or *morphological deviations* (organoid = organ-like) the tissues or organs may be normal, but their mutual relation is not. Cytological aberrations may be involved, especially in histoid deviations, as a result of improper hormonal balances. These groups merge.

A term often used in growth abnormalities is *proliferation* (L. proles = offspring or sprout; fero = to bear). It refers to uncontrolled yeast-like sprouting of cells or to uncontrolled growth and development of tissues and organs (see p. 50). Especially in other languages (Du. 'woekering', Ger. 'Wucherung') it also implies development at the cost of surrounding cells, tissues and organs. Thus it needs further specification when applied (e.g. cell or tissue proliferation = *hyperplasia*, p. 50; proliferation of sprouts and buds = witches' broom growth, p. 106).

Because in malformations, especially histoid ones, the symptoms may resemble swellings caused by parasitic organisms, the term gall has also been used in plant virology. An example is the disused name *Galla fijiensis* for Fiji disease of sugar-cane. More recently, the cecidiologist Trotter (1954) and the plant virologist Blattný (1961) used the name 'virocecidium' (L. cecidium = gall; used esp. in combination, e.g. zoocecidium = gall caused by an animal).

This must be considered inaccurate, since galls are local growth aberrations induced by animal or plant parasites intended to be used by their inhabitants as protection and source of food (e.g. Küster, 1911). Though Brakke *et al.* (1954) found that the quantity of virus in a tumour (in his case stem tumours of sweet clover) far exceeds that in normal tissues of the infected plant, its occurrence is by no means restricted to abnormal tissue and there is no evidence that it is indispensable for survival of the virus. That the virus perhaps benefits from the formation of pseudo-phloem inside the tumour (Lee & Black, 1955) does not alter the situation. The cecidiologist Docters van Leeuwen (personal discussion and 1959) agreed with me that abnormalities induced by viruses should not be classified as galls.

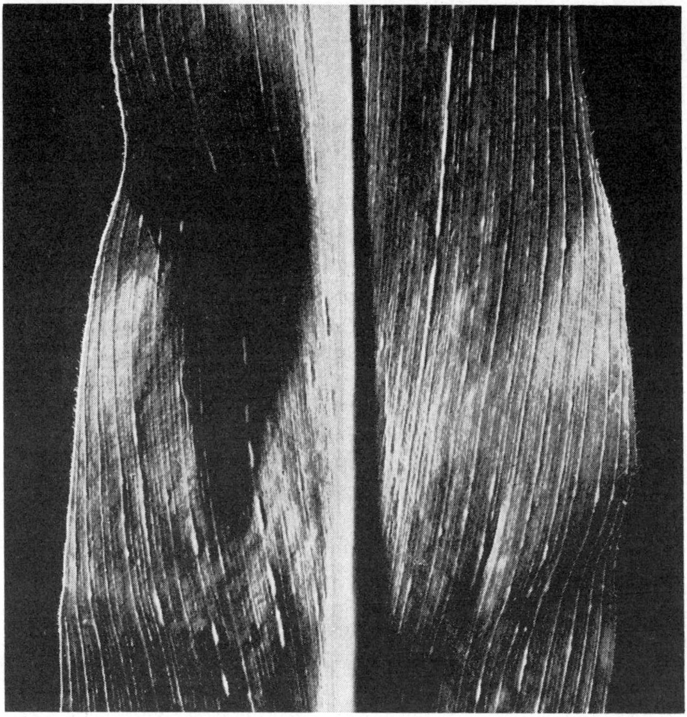

Fig. 31 Elongated histoid enations on the undersurface of a maize leaf caused by maize rough-dwarf virus. (After Grancini, 1958.)

Histoid malformations Since no normal organs are produced, *histoid malformations* cannot be described in morphological terms. Most of them are due to hyperplasia and hypertrophy.

The simplest aberrations of this category are *histoid enations* (L. e = from; natum = born, from nascere = bear, bring into existence). They are mere protuberances, peculiar protruding growths, of limited size. In sugar-cane Fiji disease the small elongated excrescences on the lower surface of the leaf veins result from abnormal proliferation of the phloem or of tissues next to it (Kunkel, 1924). A similar phenomenon occurs in rough-dwarf disease of maize (Fig. 31; Grancini, 1958), where Biraghi (1952) demonstrated the underlying proliferation of phloem tissue. Such outgrowths have been particularly studied in clover and broad bean infected with clover enation virus (Bos & Grancini, 1968). In clovers, mainly white clover, enations are usually restricted to the underside of the midrib where they are very conspicuous, being found on an often conical downward bulge of the leaf surface. The excrescences may vary in form and number from single large or small papillae or whitish spines to rows of

Fig. 32 Histoid enations in white clover *(Trifolium repens)* with a graft-transmissible virus. (After Bos & Grancini, 1968.)

irregular swellings (Fig. 32). In artificially infected broad bean plants, however, all larger veins swell irregularly causing the leaves to curl and their margins to roll upward; conspicuous chains of beady protrusions have also been found on the stems (Fig. 33 left).

Ullrich & Quantz (1964) found that the clover enations arise mainly by hyperplasia and hypertrophy of phloem and adjacent parenchyma; the phloem shows degeneration as well. Bos & Grancini (1968) have confirmed this for broad bean with the same virus, though there pericycle and cortical parenchyma are also hyperplastic. The outgrowths in broad bean often contain spindle-shaped tumours (see below; Fig. 33 right).

Clover enation disease belongs to a group of virus diseases with similar symptoms. Among them are some of economic importance, such as beet curly top and the above-mentioned maize rough dwarf.

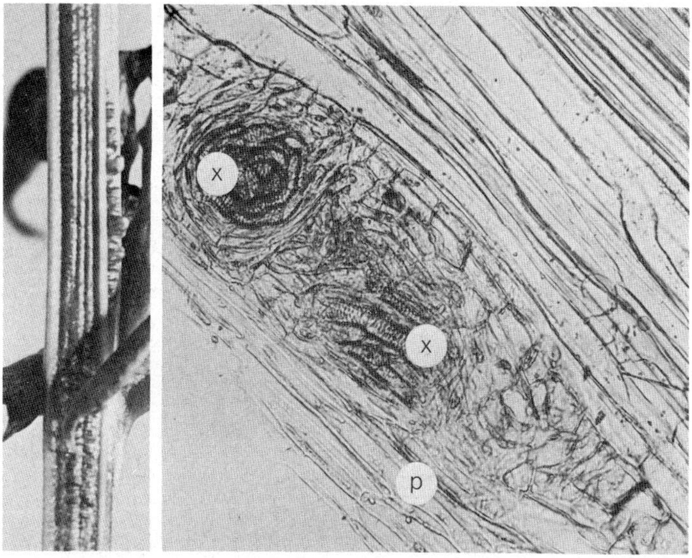

Fig. 33 Broad bean *(Vicia faba)* with chains of enations on stems (left) and spindle-shaped tumour in phloem (p) × 135 (right) after infection with the virus from enation-diseased white clover of Fig. 32; x = xylem. (After Bos & Grancini, 1968.)

Plate 7 'Black root' disease of *Phaseolus vulgaris* due to a virulent strain of bean common mosaic virus.

Plate 8 Apple rough skin in 'Schone van Boskoop'. (From F. A. van der Meer, Wageningen).

Other members are tobacco leafcurl (Indonesian kroepoek), sugar-cane Fiji disease, tobacco clubroot, rice black-streaked dwarf, citrus vein enation, and lucerne virus papillosity.

In the previous paragraphs the term *tumour* (Am. tumor; from L. tumor = swelling) has been used a few times and this indicates the absence of a sharp borderline between histoid enation and tumour. As a rule the latter is larger and more irregular, like histoid enations.

The tumour cells are abnormal only in number, in function, and sometimes in size. However, their disharmonious growth is completely autonomous and the new tissues are slightly differentiated in a manner quite unlike that of adjoining normal tissues. They can attain considerable size and develop, as 'parasites', at the cost of surrounding tissues.

Virus tumours are gaining considerable attention because they closely resemble cancers in man and animals, though there are differences because of differences in ontogeny between animal and plant tissues (Black, 1952). The study of clover wound tumour (clover big vein) in particular, has resulted in a vast literature (reviewed by Black, 1965).

Generally tumorous swellings in leaves are small. So are the enations under the veins of crimson clover (clover big vein), induced by wound tumour virus, actually produced by the development of small tumours in the veinal phloem. In petioles, stem, and roots many of them do not show on the surface as protuberances (Lee & Black, 1955). Bos & Grancini (1968) found spindle-shaped internal tumours in the swollen vascular bundles of broad bean plants infected with clover enation virus (Fig. 33 right).

Tumours induced by the wound tumour virus occur on stems and roots of sweet clover (*Melilotus albus* and *M. officinalis*; Fig. 34), and on roots of sorrel (*Rumex acetosa*) and a number of other plants (Black, 1945). On stems they may reach a centimetre in diameter. In systemically infected plants they are produced after wounding (Black, 1946). When a root starts branching, the sideroot has to break through the cortex, and this natural wounding causes tumours to develop in the pericycle close to the wounded cells (Lee, 1955). This may even occur at the base of a bacterial nodule. Woody tumours on young rough lemon and West Indian limetrees are associated with citrus vein enation virus and tissue wounding (Wallace & Drake, 1961). Their anatomy has been studied in leaf veins, thorns, and even roots

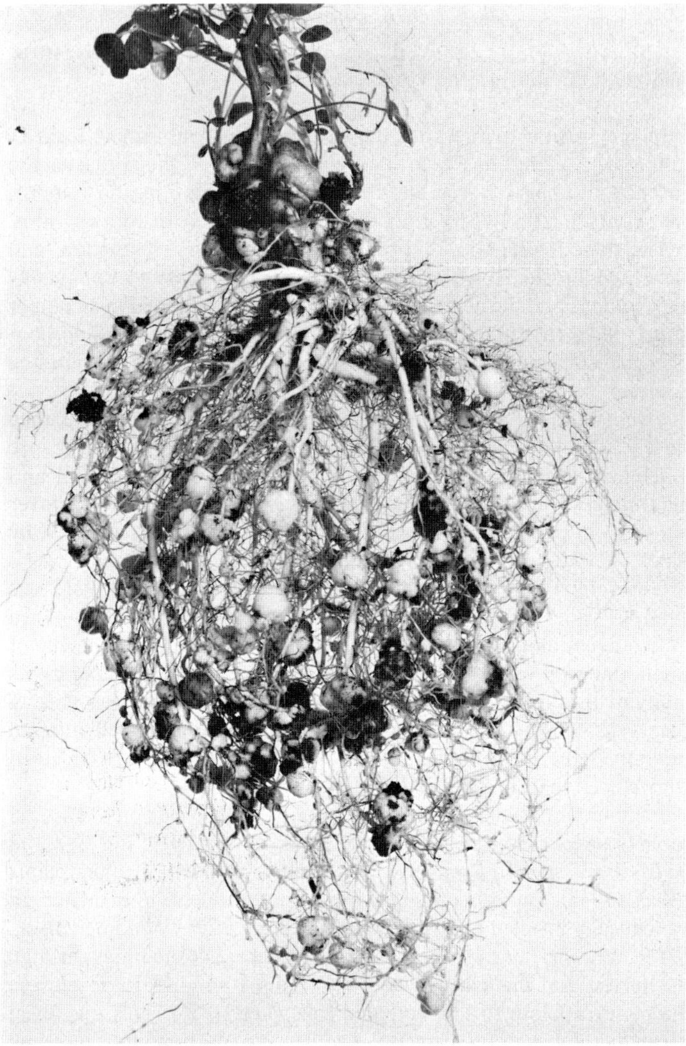

Fig. 34 Root and stem tumours in sweet clover (*Melilotus officinalis* clone C 11) after artificial infection with clover wound tumour virus. (From Dr L. M. Black, Urbana, Ill.)

by Hooper & Schneider (1969).

Histoid outgrowths may even occur on fruits, as with peach wart disease. Here bleached bumps or raised welts may develop near the top of young fruits, later stages often involving at least half of them. The warts are sometimes very hard and woody but usually tough and leathery. They form in superficial tissues; the underlying tissue is coarse and filled with gum pockets (Blodgett *et al.*, 1951).

In other cases extensive swellings develop. In stems of sweet cherry 'Napoleon' in Oregon, USA, cherry black canker virus induces

Fig. 35 Cankerous tumours in twigs of 'Napoleon' cherry *(Prunus cerasus)* with cherry black canker virus. (After Zeller *et al.*, 1951).

slightly swollen areas later splitting and growing into rough black cankers (Fig. 35; Zeller *et al.*, 1951). They are *cankerous tumours* because necrosis and secondary decay of material get involved. For swelling as a secondary reaction to necrosis in woody stems the name *tumorous canker* has been proposed (p. 82).

Even whole stems or shoots may swell, resulting in *shoot swelling*, as in the notorious swollen shoot disease of cocoa in Western Africa (Posnette, 1947). Here suckers at the base of the trunk are particularly involved, and their diameter may be twice that of a normal stem. Shoot swellings may be nodal or internodal; often they are terminal. In such cases the swelling is due to an increase in xylem tissues; the phloem swells only slightly.

In contrast to these examples of overdevelopment, certain tissues may be less developed than normal. Such a histoid deformation is *stem pitting* in apple and citrus. It is especially severe in certain varieties of crab apple but has also been observed in some other varieties

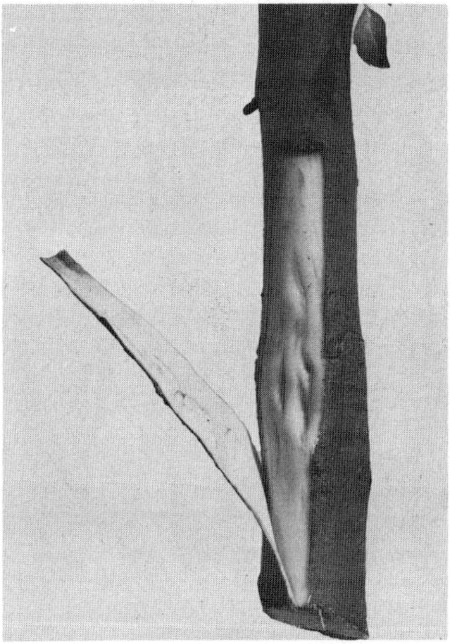

Fig. 36 Stem pitting in 'Virginia Crab' apple: pits in wood, and pegs on the inner surface of the lifted bark matching the pits in the wood. (After Posnette & Cropley, 1963.)

containing the apple stem pitting virus grafted onto 'Virginia Crab'. Pitting becomes visible on the surface of the wood when the bark is removed (Fig. 36). It consists of sparse elongated pits to numerous long, tiny furrows resulting from failure of some cambium to differentiate into normal elements, so that a wedge of phloem becomes embedded in the developing xylem. The phloem in these wedges is severely affected; so is the xylem parenchyma in the diseased wood (Hilborn *et al.*, 1965). Such wood pitting may also occur in Mexican lime *(Citrus aurantifolia)* with tristeza when localized areas of cambium degenerate and no new wood is formed (Schneider, 1959).

A related symptom is *stem grooving* of apple, caused by the apple stem grooving or E 36 virus, of which 'Virginia Crab' is also a good indicator. At the end of the first growing season young stems have a few long grooves at their base after removal of the bark. In the second year they are visible in the bark. At the end of the growing season the shoots easily break at their base (a kind of girdling effect) and some other incompatibility symptoms also occur (de Sequeira & Cropley, 1968).

A more extreme histoid underdevelopment is *branch flattening* in apple flat limb. Its first symptoms appear in two or three years old branches, which irregularly flatten (Ger. 'Flachästigkeit') or are provided with broad ribs (Ger. 'Rillenkrankheit'). With increasing thickness of the branch the furrows become deeper. The phenomenon is not an abnormal outgrowth but a reduction in growth of the xylem between the ribs, as visible in transverse sections. The results are distortions and sometimes spindly swellings (Blumer, 1956; Kegler, 1964; Scurfield & Reinganum, 1964).

Organoid malformations In organoid malformations, cells, tissues, and organs are usually almost normal, but organizational disturbances lead to deviations in the morphology of the plant. These morphological aberrations, especially those in flowers, have long attracted attention from botanists. First they were considered curiosities, but later on they gave rise to a special branch of biology, named teratology (Gr. τέρας, teras = wonder, monster). Even textbooks have been devoted to it, such as those by Masters (1869) and Penzig (1921–1922, with a wealth of literature references).

At present we know that many teratological phenomena, called *teratomata* (or, in common speech, sometimes monstrosities), are symptoms of diseases, sometimes due to viruses. These abnormalities come within the field of pathological or morbid morphology or pathomorphology (for an extensive discussion see Bos, 1957b, 1970).

The organoid malformations practically all ensue from disturbances in the supply or distribution of hormones regulating growth. That is why some virus-caused leaf malformations closely resemble those caused by an overdose of hormonal weedkillers.

Such a hormonal disturbance is suggested by many instances of *leaf narrowing*, where laminar tissue is reduced in growth, whereas midrib and main veins are almost normal. So, development of veinal and interveinal tissue is imbalanced. In various cases this laminar growth reduction is associated with discolorations like mosaic as mentioned earlier.

Leaf narrowing is striking in cherries with Pfeffinger or Eckelrader disease, where the malformation is accompanied by abnormally deep and irregular serration (Fig. 37 middle), or in yellow lupins with bean yellow mosaic virus. Very common is tomato narrow leaf caused by tobacco mosaic virus or by cucumber mosaic virus (Fig. 37 top). In tomatoes the leaflets may resemble fern leaves and of the lamina only the main veins may remain. This extreme is called *shoe stringing*, sometimes 'lacing'. Tepfer & Chessin (1959), in their study on 'Xanthi' tobacco infected with tobacco mosaic virus, mention extremes in which the 'leaf' is completely radially symmetrical with no vestige at all of mesophyll. In *Datura stramonium*, deformations caused by various strains of cucumber mosaic virus, also vary widely (Juretić, 1968). Figure 37, bottom, shows the narrowing of the palmate leaves of grapevines after infection with grapevine fanleaf virus. Here the reduction starts with deeper indentation and lobation; in a further stage the five main veins bend together as in a partially closed fan. This *fanleaf formation* gave the name fanleaf disease.

In plants with flowers that normally are tubulous or gamophyllous in any other way, leaf narrowing may lead to loose-leaf flowers (or split corollas) as in tobacco plants with tobacco streak virus (Costa & Carvalho, 1961; personal observation).

In contrast to the just-mentioned leaf abnormalities, some others result from less growth of the veinal tissue relative to laminar tissues. This leads to a lumpy bubbly leaf surface as in tobacco leafcurl, with sunken veins and elevated interveinal tissue, termed *rugosity*. When the deformation is more irregular, and is associated with furrowing and wrinkling of the lamina, the terms *crinkling* and *curling* are used,

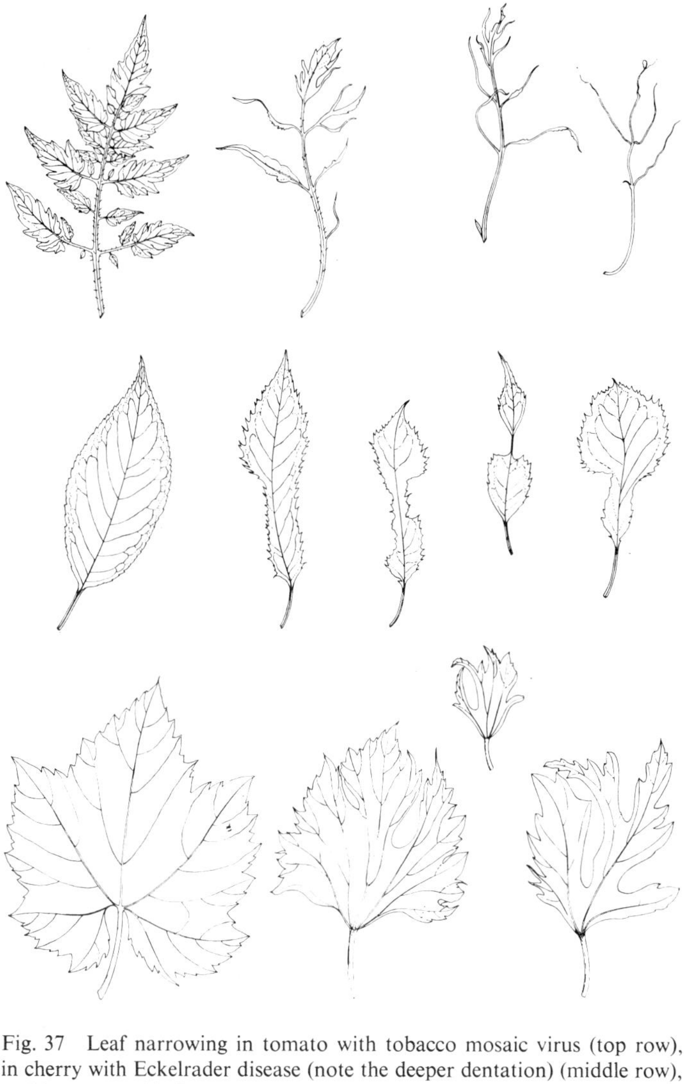

Fig. 37 Leaf narrowing in tomato with tobacco mosaic virus (top row), in cherry with Eckelrader disease (note the deeper dentation) (middle row), and in grape with fanleaf disease (bottom row); healthy leaves, left. (By Miss I. Zewald, Dept. Plant Taxonomy and Plant Geography, Agric. Univ., Wageningen.)

the latter indicating a more irregular and sometimes contortive type of deformation. Examples are turnip crinkle and sugar-beet leaf curl. Such phenomena, especially the more irregular ones, may also be secondary, due to anatomical or other disorders of veins.

Another imbalanced development is reduction of petiole and midrib, or rachis of a compound leaf, typical of potato bouquet disease (Köhler, 1952), where it leads to crowding of leaflets and sickle-shaped downcurving of their midribs.

Epinasty (Gr. ἐπί, epi = on, upon; ναστός, nastos = pressed close), the more rapid growth of the surface of an organ, such as the petiole or the leaf blade, is quite common in virus-infected plants. It leads to downcurling of the entire leaf. It is not known whether regular upcurling as in potato leafroll, or the downcurling over the whole length of the leaf, both known as *leafrolling*, are similar in mechanism to epinasty.

A peculiar morphological aberration is considerable enlargement or hypertrophy of stipules of apple leaves with witches' broom disease. Here the stipules are dentate and clearly differentiated into petiole and lamina. The symptom has a high diagnostic value, but that the pathogen is a virus has been doubted (p. 104).

Enations described earlier were histoid deviations, but some others on leaves are organoid, because internally they have the structure of a normal leaf. Such *organoid enations* usually develop under the leaf. They have been studied, for instance, with common tobacco mosaic virus in *Nicotiana paniculata* and *N. tomentosa* (Jensen, 1933), for tobacco leafcurl virus ('kroepoek' virus) in tobacco (Kerling, 1933), for pea enation mosaic virus in pea, broad bean and crimson clover (McWhorter, 1950), for Pfeffinger or Eckelrader disease in cherry (Stoll, 1952), for tomato aspermy strain of cucumber mosaic virus in leaves of *Nicotiana glutinosa* and *Lycopersicum esculentum* (Praceus, 1958), and even in the corolla of *Petunia hybrida* (Noordam, 1952). The enations may occur between veins or near them, rarely also on veins, as with tomato aspermy virus, or even restricted to them, as with Eckelrader disease or tobacco leafcurl. With pea enation mosaic they usually surround chlorotic areas in the mesophyll as fringes or sometimes merely as rough ridges. Often, however, the enations are leafy or look like wings, cups, boats, funnels or shells.

The anatomy of organoid enations has been studied by Kerling (1933) for tobacco leafcurl virus, by Praceus (1958) for tomato

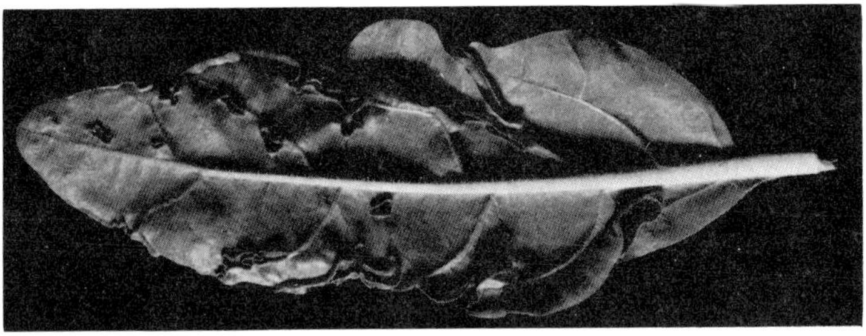

Fig. 38 Organoid enations in wild tobacco *(Nicotiana rustica)* with tomato aspermy virus. (After Kristensen & Thomson, 1958.)

aspermy virus (Fig. 38), and by Ullrich & Quantz (1964) for pea enation mosaic virus. In all of them the leafy enations contain an upper epidermis, a palisade layer, spongy parenchyma, and a lower epidermis as normal. When they start as cup-like structures around chlorotic areas, their upper epidermis is on the inside, the outside epidermis being directly connected with the lower epidermis of the leaf. Their origin around chlorotic areas suggests a loss by these areas of their original morphogenetic control over the surrounding tissues.

Misshapen stems also may be due to an unbalanced development, growth lagging behind in certain tissues only. Virus-diseased plants may show *internode shortening*. Branches of grapevine affected by fanleaf disease usually have short internodes, or some internodes may even be absent, leading to '*double nodes*' (Hewitt & Gifford, 1956). When all internodes of a stem are reduced the leaves become rosetted. Infections late in plant development may result in a crowding of the leaves at the branch tips. Examples of virus diseases characterized by this *rosetting* are apple rosette, peach rosette, Pfeffinger or Eckelrader disease of sweet cherry (in German sometimes called 'Rosettenbüschelkrankheit') and groundnut rosette. The bushy appearance may sometimes be confused with witches' broom formation, but this is an entirely different symptom. Usually in rosette diseases the leaves involved also show some abnormalities.

Another morphological stem disorder is the peculiar *zigzag growth* of grape canes in vines with fanleaf. Together with other malformations, such as short internodes and double nodes, this phenomenon may be of diagnostic value when the vines are dormant and have no leaves (Hewitt & Gifford, 1956).

A very peculiar effect of a virus in the morphology of fruits is the partial or complete suppression of spines on the capsules of thorn-apple *(Datura stramonium)*, as reported to be caused by *Datura* quercina virus (Blakeslee, 1921), tobacco etch virus (Kunkel, 1944) and cucumber mosaic virus (Aubert, 1960; Juretić, 1968).

Witches' broom phenomena

Witches' broom phenomena, though organoid in structure, constitute a special category of growth abnormalities. They are reminiscent of witches' brooms caused by local infection of certain parasites or local action of growth-stimulating substances excreted by such parasites (Bos, 1963), but the phenomena to be dealt with here differ in being due to systemic infections which modify the growth of the entire plant.

I have extensively studied the patho-morphology of witches' broom phenomena (Bos, 1957a). Under 'witches' broom virus diseases' I brought together numerous important leafhopper-borne diseases such as the North American aster yellows, the Australian tomato big bud, the East European stolbur (mainly of Solanaceae), the West European *Rubus* stunt, and the almost worldwide clover phyllody or virescence. Additional data have been surveyed by Bos & Grancini (1965).

The name 'yellows-type diseases' currently often used for this group of disorders (Kunkel, 1954b; Valenta *et al.*, 1961) is inadequate because in aster yellows, the most studied representative of the group, chlorosis and yellowing are not invariable symptoms (Kunkel, 1931), and because the name suggests a relationship with sugar-beet yellows and some other persistent aphid-borne virus diseases.

Witches' broom diseases, long ascribed to viruses, are now increasingly ascribed to mycoplasma-like organisms. Much about their nature is still unknown, but in many ways they indeed behave like viruses, so that inclusion in this book remains justified. For simplicity, the incitants are still often denoted as 'viruses'.

As early as 1896, Hugo de Vries reported an 'epidemic of virescences' in 27 species of plants in his garden in 1893 and 1894. Though his careful macroscopic and microscopic search for parasitic organisms was in vain, he was soon convinced of the disease's infectivity and even of its distribution by flying insects. He wrote 'to hope and trust that in this, others may be more successful and I flatter myself with the thought that the knowledge of the facts to be reported will make the way somewhat easier' (see Bos, 1957b, 1966).

Only in 1924 did Kunkel (1924a) prove the infectious nature of a witches' broom disease (aster yellows) and its transmission by a leafhopper. Since then, viruses were held responsible, for lack of visible organisms, because of the systemic nature of infection, because of association with other virus symptoms (vein chlorosis, chlorosis, phloem degeneration), and because of vector relationships characteristic of various viruses.

Over 50 plant diseases are now associated with mycoplasma-like pleomorphic minute filterable organisms (Fig. 58); e.g. Ploaie & Maramorosch, 1969). Doi *et al.* (1967) first reported such organisms from Japan for mulberry dwarf, potato witches' broom, *Paulownia* witches' broom and aster yellows. The organisms can be detected by electron microscopy in phloem of infected plants and in their vectors after ultrathin sectioning. Transmission and symptoms are affected by certain antibiotics (Davis *et al.*, 1968). Evidence that these micro-organisms are actually incitants of witches' broom diseases is thus increasing, though Koch's postulates are still hard to apply. Even if such proof becomes possible, the pathogens behave so much like viruses that their study may remain the domain of virologists.

Fig. 39 Witches' broom growth in bilberry *(Vaccinium myrtillus)* naturally infected (right). Unlike the healthy plant (left) the diseased one kept its foliage in the greenhouse through the winter. (After Bos, 1960.)

Witches' brooms caused for instance by certain fungi or mites consist of a local mass of slender branches. All the buds develop into sprouts (even those that normally remain dormant), growth of newly formed branches is abnormally upright and flowering is completely absent.

If plants are infected early in growth, systemic infection by witches' broom 'virus' results in an entirely abnormal plant, as with whortleberry *(Vaccinium myrtillus*; Fig. 39; Bos, 1960). Such plants show *witches' broom growth*; the erect habit of new branches indicates *intensified negative geotropism* and contributes to the bushy habit. They rarely flower. For excessive development of sprouts the terms *blastomania* and *cladomania* are used (Gr. βλαστός, blastos = sprout, shoot; κλάδος, klados = branch; μανία, mania = madness). The term proliferation is used in apple proliferation disease (Mulder, 1953), but is not specific enough. Here it refers to proliferation of branches (see p. 91).

After infection at an early stage, a plant may remain small and resemble a dense bushy broom, as in bunchy plant of *Arachis* (groundnut) (Morwood, 1954). This abnormality is often called 'stunting', as in *Rubus* stunt (Prentice, 1950a), but real stunting is quite different.

Internal spread in woody plants may be slow when the pathogen enters an older specimen, and for a long time the infection may remain local. The term *witches' broom formation* refers to such local brooms, as in bunch disease of pecans (*Carya pecan*; Cole, 1937) and in brooming disease of black locust (*Robinia pseudoacacia*; Hartley & Haasis, 1929) also. Plants with witches' broom disease usually have small leaves (Fig. 39), as in little leaf of brinjal *(Solanum melongena)* in India (Thomas & Krishnaswami, 1939), and in legume little leaf (Hutton & Grylls, 1956) for a similar disease of subtropical pasture plants in Australia. Since small leaves are a common symptom of virus diseases, the name is not indicative of witches' broom.

Infection of older plants may hardly change their general appearance. So in the Australian lucerne witches' broom disease, sometimes called bunchy top (Edwards, 1935), the only symptom is abnormal leafy branching in parts normally developing into inflorescences (Fig. 40). Upon closer observation such terminal brooms turn out to consist of modified flowers (cf. the name cranberry false blossom; Shear, 1916). I have given detailed descriptions (Bos, 1957a) for *Crotalaria* spp. collected in Indonesia and for *Tropaeolum majus* found at Wageningen. In *T. majus* the abnormality is more pronounced

Fig. 40 Greening and phyllody, notably of carpels, in inflorescences of *Crotalaria usaramoensis*; the right one was younger than the left when infected. For details see Fig. 42.

if the flowers are younger at infection, and I found a series of abnormalities progressing from a normal flower to a completely vegetative leafy branch (Fig. 41). This metamorphosis of floral parts is called *antholysis*, a term derived from teratology (Gr. ἄνθος, anthos = flower; λύσις, lysis = loosing, dissolution). It entails virescence, phyllody, and prolification, and it indicates that the pathogen completely, and usually suddenly, causes the arrest of flower formation, even where it had already started in the bud. Buds partly initiated as flowers usually develop vegetatively according to stage of initiation at infection.

Antholysis starts with *virescence* or *greening* (L. virescere = to become green) of normally white or coloured floral parts by the development of chloroplasts. It is characteristic of such diseases as virescence of tomato, a synonym for tomato big bud (Hill, 1943), and green flowering disease of *Justicia gendarussa* (Su, 1933) and *Sesamum indicum* (Robertson, 1928), of green petal of strawberry (Posnette, 1953), and of clover virescence (Plate 9).

When, in later stages of antholysis, floral parts develop as almost normal foliage, the phenomenon is called *phyllody* (Gr. φύλλον, phyllon = leaf). In sepals this is often associated with considerable hypertrophy, as in raspberry flowers infected with *Rubus* witches' broom (*Rubus* stunt) 'virus'. In tomatoes infected with stolbur or tomato big-bud 'virus', the calyx enlarges considerably, even before flowering, leading to big (flower) bud symptom. Phylloid stamens are rather rare, but they may occur, as in *Tropaeolum* (Fig. 41E;Bos, 1957a). Phyllody of carpels is very striking: the enlarged more or less open carpels usually protrude beyond the virescent corolla (Fig. 40, 41B-F, 42: *pi* and *c*). Ovules can be changed into small leaf-like organs at the edge of the open carpel (Fig. 42 *o*). On account of such phenomena 'virus' diseases sometimes have been named phyllody, such as phyllody of sunn hemp *(Crotalaria juncea)* (Bose & Misra, 1938) and phyllody of *Sesamum indicum* (Pal & Nath, 1935), both in India.

The next step in antholysis is *apostasis*, the development of the internodes theoretically present in the floral receptacle (Gr. ἀπό, apo = away from, without; στάσις, stasis = a standing still, arrest of motion or action). In other cases only the gynophore is involved (Figs. 41F and 42*g*) so that the also abnormal gynaecium extends beyond the flower, contributing to the protrusion of the phylloid carpels as mentioned above.

The elongation of the receptacle above the insertion of the pistil

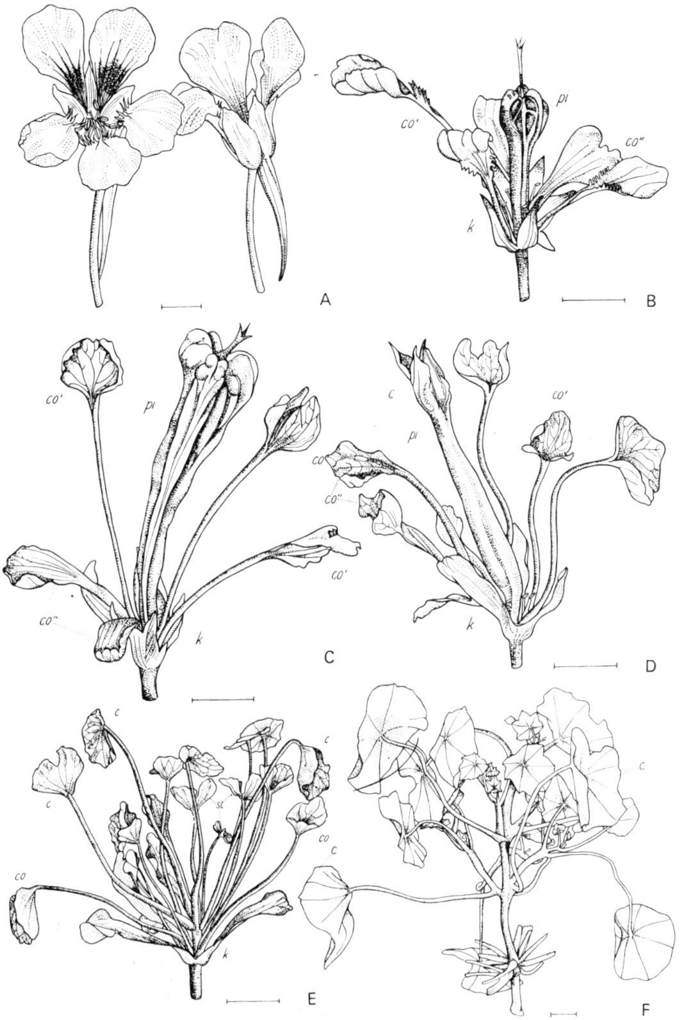

Fig. 41 Antholysis in nasturtium *(Tropaeolum majus)* with virescence and phyllody increasing from B-F; c = calyx, co = corolla, st = stamens, pi = pistil composed of three carpels (c). In F, from which calyx, corolla and stames are omitted, prolification in the centre of the flower and from carpel axils is striking. For various magnifications see bars representing 1 cm. (Modified from Bos, 1957a.) For A, B and C see also Plate 9b.

and its development into a sprout is quite common. In teratology this phenomenon is called *prolification*, meaning the production of a sprout or sprouts (L. proles = offspring, sprout; facere = to make; Figs 41F and 42: *p*). When it involves the apex of the receptacle, the term *central prolification* is applied, whereas *lateral prolification* refers to sprouting from the axils of the floral organs (Fig. 42: *p*').

As well as prolification the word proliferation has been used, which means bearing sprouts. Since in literature this term has also been applied to uncontrolled sprouting of cells resulting in an almost unlimited development of tissues (p. 91), the term prolification is preferred here for the 'sprouting' of flowers.

The new sprouts are usually vegetative. They often grow into leafy branches, but a new inflorescence may also develop, especially after central prolification, though usually with abnormal flowers. The vegetative type of floral prolification is homologous with witches' broom growth in vegetative parts of a plant and greatly contributes to its bushy growth.

Finally entirely vegetative sprouts instead of flowers may occur, e.g. in the axil of a bract in the upper part of an abnormal inflorescence.

Witches' broom phenomena may lead to complete *sterility*, as indicated by the name sterility of *Sesamum indicum* in India (Roy, 1931), resulting in important economic losses, as infected plants do not produce fruits or seeds.

As prolification in flowers greatly contributes to witches' broom growth, the term *witches' broom phenomena* has been introduced (Bos, 1957a) for the syndrome comprising both witches' broom growth in the vegetative parts of the plant and the complex of antholysis in the sexual parts. Which symptoms dominate depends on circumstances, on plant species, and largely on stage of growth at infection. Witches' broom growth is most striking after infection during vegetative development; after infection during flower initiation, floral abnormalities prevail. The differences in terms might suggest non-existent differences in disease.

Some special deviations accompanying witches' broom phenomena should be mentioned here. In *Phaseolus calcaratus*, seeds may germinate in immature pods formed before infection (Bos, 1957a), suggesting *reduction of seed dormancy*. Another striking example of 'vivipary' is in tomato fruits formed before infection by potato witches' broom (Limberk & Ulrychová, 1972). Such a *disturbed periodicity* also induces the continuation of growth of plants diseased

Plate 9 Virescence and phyllody in white clover. Note the phylloid carpels protruding beyond the green flowers.

Plate 10 Virescence and phyllody in *Tropaeolum majus* with witches' broom disease; see also Fig. 41A, B and C. (From Bos, 1957a).

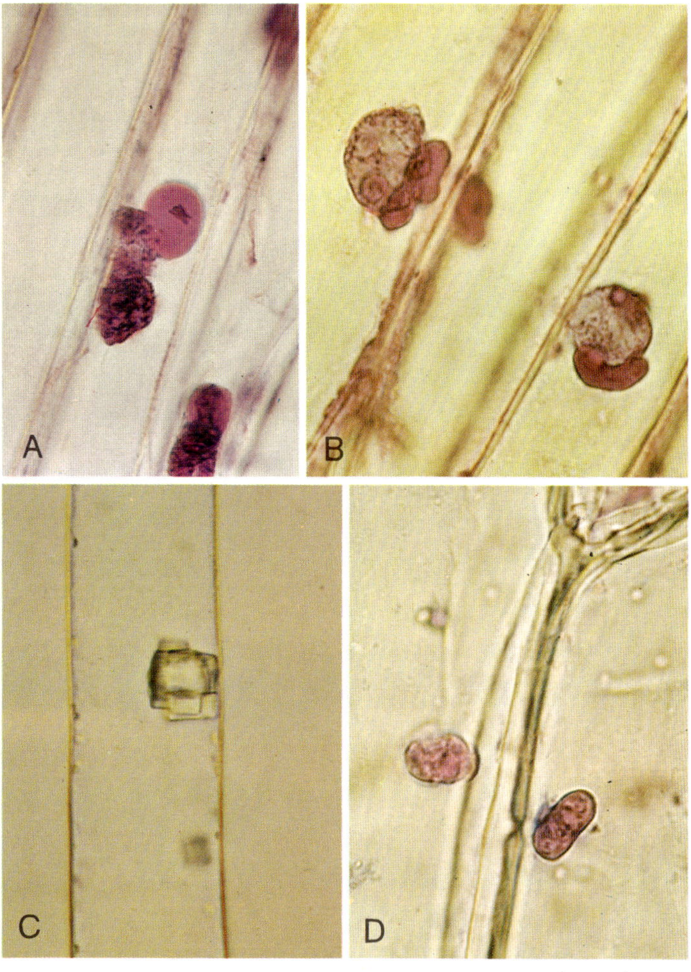

Plate 11 Inclusion bodies in epidermal cells from petioles (A, B, D) and in hair cell (C), stained with phloxine and methylene blue (A, B, D) and unstained (C). A, Broad bean with bean yellow mosaic virus; B, and D, radish (B) and white cabbage (D) with cauliflower mosaic virus; C, tobacco hair cell with tobacco mosaic virus from tomato. Magnification: A 600×; B 1000×; C 300×; D 1000×.

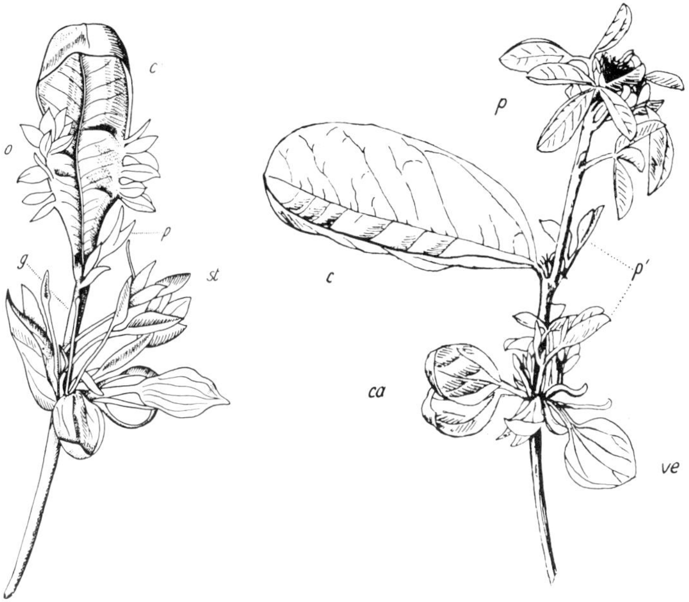

Fig. 42 Two virescent flowers of *Crotalaria usaramoensis* with striking phyllody of carpel (c) and ovules (o). Note also the central (p) and lateral (p') prolification and the development of a gynophore (g) or stipe; (left, × 1,8; right, × 1,3). (After Bos, 1957a.)

with witches' broom during autumn and its early start in spring, as observed in *Vaccinium myrtillus* (Bos, 1960). Another example of disturbed periodicity is carrot plants flowering the first year (although with abnormal flowers) instead of the second year, after infection with a witches' broom 'virus' of clover (Fig. 43; Bos & Grancini, 1965).

In *Arachis hypogaea* with normally downgrowing stipes (stalk-like prolongation of the receptacle beneath the ovary) meant to bury the developing pod in the soil, *intensification of negative geotropism* due to infection causes the stipes to curl upward again, even when they had already penetrated the soil and started pod formation. The same phenomenon can be observed in tomato plants with stolbur or tomato big bud, and in cranberry with false blossom (Shear, 1916), where the pedicels, instead of bending down, remain upright.

Fig. 43 Witches' broom growth and bolting in young carrot plants after artificial infection with clover phyllody (right). (After Bos & Grancini, 1965.)

Infection with the potato witches' broom 'virus' may produce an exceptionally large number of small tubers in potatoes (Todd, 1958) or even *aerial tuber formation*, the latter also with aster yellows (Severin, 1940) and stolbur (for more literature see Klinkowski, 1958). But aerial tuber formation may also be a secondary effect of disturbed translocation of carbohydrates caused by phloem degeneration. Tubers of potato with stolbur (Savulescu & Pop, 1956), potato witches' broom (Todd, 1958), aster yellows (Larson, 1959), or tomato big bud (Webb & Schultz, 1958) often show *spindle sprouting* or *hair sprouting* after germination, and produce spindling, spindly or hair sprouts.

In summary, witches' broom diseases are mainly characterized by (1) reduced apical dominance leading to excessive outgrowth of buds, (2) enhanced negative geotropism, sometimes even a reversal of positive geotropism, and (3) suppression or mostly even complete prevention of sexual development. The processes are generally accepted to be controlled by hormones, and disturbed hormonal balance is evident.

According to type of witches' broom disease or strain of pathogen, to host plant and conditions, more or less severe phloem degeneration may be involved as seen in aster yellows (Rasa & Esau, 1961) (see also p. 53). In turn this may lead to transport disturbances, yellowing, growth reduction, stunting, and other secondary deviations complicating or sometimes even predominating over the witches' broom syndrome. Some other diseases, like pear decline and elm phloem necrosis, more recently attributed to mycoplasma-like organisms, are characterized by deviations resulting from phloem degeneration. Symptomatologically the latter diseases are similar to those caused by certain phloem-limited viruses, especially the true yellows-type virus diseases or yellowing diseases.

There is a certain relationship between witches' broom growth and tumour formation. In tumours development of new tissues with low internal organization is completely autonomous. With witches' broom, growth is also nearly unlimited, but the new tissues exhibit a quite high internal organization, though lower than normal, as apparent from the absence of flowers.

Miscellaneous abnormalities

Finally, various externally visible abnormalities resulting from virus infection are known which are hard to assign to the previous categories of this chapter. The more important ones will be treated in this section.

Interesting is the effect of tobacco mosaic virus on the growth rhythm of the leaves of 'Samsun' tobacco, as revealed by time-lapse cinematography (Novák, 1964, 1975). Leaf movements are chaotic for ca. 12 days during the period of rapid virus multiplication beginning some 4 days after inoculation. This period is followed by stagnation of movement and growth for about 8 days. Thereafter, another period of chaotic movement for about 8 days occurred and this was finally succeeded by a return to the rhythm of normal growth.

Some diseases may advance flowering; others may retard it; others again may cause premature dropping of flowers and fruits (examples in Holmes, 1964). Fruit texture and flavour are often abnormal through aberrations like necrosis.

Premature *leaf dropping (leaf abscission, leaf casting, leaf fall, defoliation)* may be a disturbance of correlation and is rather common in virus diseases. It is very striking in sour cherry with cherry yellows (Keitt & Clayton, 1943) where the older leaves start falling, even before any chlorosis is visible; 50% of the leaves may get involved. With *Capsicum annuum* leaf casting has been reported as an effect of hypersensitivity to local infection with tobacco mosaic virus (see Holmes, 1964). 'Leaf dropping' due to potato virus Y in some potato varieties (sometimes called potato leaf drop streak) is not a leaf abscission but a drooping: it is associated with wilting and withering of the leaf; the hanging position of the leaf is primarily due to necrosis extending down the petiole.

Reduced vigour usually shortens the life of virus-diseased plants, but sometimes vegetative growth continues after the normal growing season, especially in plants with witches' broom growth (presumably caused by mycoplasma-like organisms) as in *Vaccinium myrtillus* (Bos, 1960). In the greenhouse, potato plants with witches' broom disease produce tubers, but continue vegetative growth for successive years (see Holmes, 1964). Bos (1957a) has succeeded in maintaining for over 14 years vegetative growth of *Tropaeolum majus* diseased with witches' broom, a plant which is normally an annual.

In some diseases, seed content in fruits is considerably reduced. Seed formation is almost completely suppressed in tomato fruits set after infection with tomato aspermy virus (Blencowe & Caldwell, 1949); reduced pollen and egg cell viability are both involved.

Sometimes the virus affects pollen more than ovules. *Pollen sterility* may be almost general in *Datura stramonium* with quercina virus (Blakeslee, 1921); this sterility also occurs with tobacco ringspot virus (Valleau, 1932), barley stripe mosaic virus (Inouye, 1962) and lettuce mosaic virus (Ryder, 1964). In some raspberry clones there are high percentages of shrinkage and *abortion of pollen*, as judged by staining with acetocarmine; it may be caused by black raspberry necrosis, raspberry mosaic, tomato ringspot and raspberry vein chlorosis viruses (Freeman *et al.*, 1969). Edwardson & Corbett (1961) found *male sterility*, the absence of viable pollen, to be graft-transmissible in petunia. According to Atanasoff (1964, 1971) there are good reasons to suppose that all cases of male sterility previously ascribed to cytoplasmic heredity are the result of visible or latent virus infection.

In *seed abortion* developing after infection, the role of necrosis is not completely understood. In beans with pod mottle, it involves embryos or whole ovules (Zaumayer & Thomas, 1948), as in citrus with stubborn disease (Carpenter *et al.*, 1965) and in sweet cherry with necrotic ringspot virus (Attafuah, 1965). With some viruses it may lead to reduced drupelet set in raspberry and to fruit malformation, known as crumbly fruit (p. 91).

An interesting effect of virus infection on the ratio of male and female flowers in cucumber is mentioned by van Koot & van Dorst (1958) with cucumber mosaic virus where the number of male flowers is doubled and that of female flowers halved. Cucumber green-mottle virus has the opposite effect, though less marked. Presumably a physiologically disturbed balance can be easily brought about, because the changes depend largely on temperature and daylength.

In some diseases *abnormal secretion* is characteristic. In 'Shirofugen' flowering cherry, widely used for testing stonefruit trees for ringspot viruses by bud-grafting, it is even diagnostic. Infection shows up, four to six weeks after grafting, as a necrosis around the inserted bud, accompanied by gum flow (Fig. 44; Milbrath & Zeller, 1945; Mischke, 1966). In bark of stems, branches and twigs of apricot trees infected with a stonefruit ringspot virus gum exudes even more profusely and causes boils up to 50 cm in diameter to form (Pine, 1963). In both diseases this excessive exudation may result from extreme internal gummosis associated with necrosis and degeneration of tissue (p. 50).

Several viruses impede grafting and budding in fruit-trees and thus increase *graft incompatibility*. This is a highly complex phenomenon, ascribed to various factors, such as differences in seasonal period of growth and in vegetative vigour, biochemical differences, mechanical blockage at the union, different rates of callus formation and viruses (Herrero, 1951); Hartmann & Kester, 1968). All such factors may be physiological or mechanical, and of course viruses can easily impair the equilibrium between stock and scion needed for successful union. That is why virus-induced growth reduction has been mentioned earlier in this book as a possible non-specific cause of incompatibility.

Toxopeus (1936) first pointed to an incompatibility in citrus between sweet orange scions *(Citrus sinensis)* and sour orange stocks *(Citrus aurantium)*. He excluded mere physiological factors, because the use of compatible intermediates neither prevented nor delayed the disorder. He concluded that the scion produced some systemic substance toxic to the stock. The abnormality is now known to be due to citrus tristeza virus. When present in the scion, it induces phloem necrosis in the hypersensitive rootstocks, thus creating a blockage.

Fig. 44 Excessive gum exudation in 'Shirofugen' flowering cherry after bud grafting from fruit trees with cherry necrotic ringspot virus. (After Mischke, 1966.)

In later infection, in a well developed tree, phloem necrosis below the bud union causes a sort of delayed incompatibility or bud-union disease, resulting in progressive disturbances of internal transport and associated phenomena. These finally lead to slow or quick decline syndromes in citrus tristeza and pear decline, discussed in the section on sequence of symptoms (Schneider, 1959; Batjer & Schneider, 1960).

Schossig *et al.*, (1965) demonstrated in extensive trials that infection of pear scions with ring-pattern mosaic virus budded onto pear seedlings, of sour cherry with *Prunus* necrotic ringspot budded onto *Prunus mahaleb*, and of plum with plum line-pattern virus budded onto *Prunus myrobalana* considerably reduces the chance of the bud taking.

Viruses, even if latent, are now generally accepted to play a role in incompatibility. For more examples, see Mosse (1962) and Marenaud (1966).

Increased predisposition to non-viral diseases

Virus infection, even if latent, is usually accompanied by a general reduction in vigour. This may decrease plant size and yield (see growth reduction), or weaken the plant physiologically, so that symptoms become evident only under unfavourable conditions, such as drought or frost. Thus 'Ladino' white clover infected with alfalfa mosaic virus seems to suffer from frost more than healthy ones (Roberts, 1956), and red clover behaves more like a biennial than perennial, when infected by some viruses, because of excessive winter kill (Oshima & Kernkamp, 1957).

The general weakening by virus infection may also manifest itself in an *increased susceptibility to secondary infections* with fungi and bacteria, or in *increased predisposition to other diseases.* A few examples may illustrate this.

Top yellows in pea and broad bean had long been ascribed to the soil-inhabiting *Fusarium solani*, because this fungus can commonly be isolated from diseased plants. It is now known that it may occur, though superficially, in healthy plant roots too, and that it penetrates deeper under adverse conditions for growth, in particular when the plants are infected by the persistent bean leafroll virus (Hubbeling, 1954; de Fluiter & Hubbeling, 1955). The virus causes phloem necrosis which evokes the yellowing and also disturbs transport of carbohydrates and weakens the root system.

More recently, Farley & Lockwood (1964) found that three cultivars of pea under a wide range of greenhouse conditions are more susceptible to root rots caused by *Aphanomyces euteiches* and *Fusarium solani* f. *pisi* after infection with any of four important viruses. Root rot and plant decline in red clover, caused by *Fusarium oxysporum*, *F. roseum* and *Tetracocosporium paxianum*, develop only in plants infected with clover yellow mosaic virus (Watson & Guthrie, 1964). Later Beute & Lockwood (1968) pointed to the fact that with the two first-mentioned fungi susceptibility is not increased, but that the inoculum potential of the pathogens around the virus-infected roots is higher because of increased exudation of nutrients from the roots.

Similarly, leaves of broad bean plants infected with bean leafroll virus were more susceptible to *Botrytis fabae* (Tinsley, 1959). To ascribe this to weakening of the plant by the virus seems also an oversimplification, as an increased carbohydrate content of the virus-infected leaves may make them more nutritious for the fungus. The reverse may also occur: grapevine with grapevine leafroll virus is more resistant to powdery mildew *(Uncinula necator)* than a healthy plant (Goheen & Schnathorst, 1961).

In recent years susceptibility to fungi and bacteria with previous virus infection have come to notice. Peach trees with various viruses show more secondary infections with the fungi *Cytospora leucostoma* and *Coryneum beijerinckii*, and with *Bacterium tumefaciens* than healthy plants (Corte & Montemartini, 1965). In greenhouse trials, Mwanza & Williams (1966) showed that maize and wheat with maize dwarf mosaic virus and a virus resembling wheat streak mosaic virus were more susceptible to various pathogens causing root rots, stalk rots and seedling blights.

Sugar-beet infected with sugar-beet yellows virus is more severely affected by leaf spot due to *Cercospora beticola* (literature reviewed by Bennett, 1960). Bennett assumes that this virus also increases losses from *Rhizoctonia* root rot and certain other soil-borne organisms and that secondary damage of this type may sometimes be more severe than the direct effect of yellows. In beet yellows often two viruses are involved: Russell (1968) found that infection with beet mild yellowing virus predisposes the beet to attacks by *Alternaria* spp. and *Erysiphe betae*, whereas infection with beet yellows virus has little effect on susceptibility to *Alternaria* but increases resistance to *E. betae*. With some isolates of either virus, susceptibility to downy mildew *(Peronospora farinosa)* greatly increased in some, but not all, breeding lines of beet.

Virus infection may also make plants more palatable to insects or mites, thus even influencing the epidemiology of the virus disease. So Thresh (1967) reported that reversion-diseased blackcurrant bushes are considerably more infested by the virus vector, the mite *Phytoptus ribis*, causing rounded bud galls.

Mechanical barriers that normally prevent attacks by parasitic organisms, are sometimes disturbed by virus infection, so that plants may be attacked more by secondary pathogens. Kegler (1964b) noted that bark splitting, associated with flat limb in apple, or frost-damaged spots, provide points of entry for fungal parasites such as *Nectria galligena*. Both may cleave the bark up to the wood and sometimes cause complete decline of the tree.

Virus diseases are often accompanied by *rotting*, but this is never caused by the virus itself, as death of host cells halts virus activity. Rotting or decay of dead tissue is due to the secondary action of certain saprophytic fungi or bacteria invading the weakened, dying, or necrotic plant tissues. Bacteria cause a decomposition called wet rot; fungi usually cause a dry rot, as in stolbur-affected potato plants, which are more susceptible to *Colletotrichum atramentarium* (Wenzl, 1964). Secondary infection of leafroll-infected peas with *Fusarium solani* already mentioned, leads to foot rot. In persistent necrotic lesions, as in woody stems, secondary rotting may produce an ulcer-like effect, as often occurs in cankers.

These examples again demonstrate the complexity of disease. Predisposition (susceptibility and sensitivity) to a given pathogen depends largely on host physiology. This in turn is greatly influenced by infection. Here viruses, frequently not killing the host or even resulting in latent infection but still influencing host physiology, play an important role. I can only mention in passing how susceptibility to certain viruses may greatly increase by or sometimes even depends on infection with other viruses or other pathogens, and how multiplication of one virus may be greatly enhanced by infection with another one. Through increased physiological or mechanical susceptibility, other viruses and parasitic organisms may thus enter the picture, further complicating the original syndrome.

Virus particles and inclusion bodies

So far, this book has dealt almost exclusively with the effects of viruses on their hosts, i.e. host reaction. In the past, plant virology has been equally one-sided, because the viruses themselves could long not be observed. Perceptibility of the disease agent was an essential part of the definition of infectious disease as postulated by Koch. Ever since Mayer (1886) proved that tobacco mosaic was infectious in absence of a visible incitant, viruses have been an anomaly in the definition of pathogen. They could not be removed by mechanical means such as filtering and they could not be seen. Whereas live pathogenic fungi are visible as moulds or fruiting bodies, and bacteria as visible colonies on or in the host, viruses were assumed to play no part in the visible syndrome.

This has all changed with the advent of the electron microscope. The hazard of attributing a disease to virus infection without seeing the virus particles is demonstrated by the recent data on aster yellows and the related witches' broom diseases. Since 1967, such diseases have been increasingly found to be associated with minute mycoplasma-like organisms rather than with viruses (p. 105). Observation of virus particles in sap, tissues and cells of the host is now an essential part of plant virology.

Visualization of virus particles

In 1939, Kausche, Pfankuch & Ruska first observed virus particles (of tobacco mosaic) in the electron microscope, thus directly demonstrating their corpuscular nature. Since then, shape and size of many viruses have been studied.

Shadow-casting (Williams & Wyckoff, 1945) and especially negative staining (Brenner & Horne, 1959) of virus suspensions have greatly facilitated the observation of structural details. Brandes' very simple dipping method (1957), and an even simpler combination with negative staining (Hitchborn & Hills, 1965) of crude plant sap allow rapid and more direct demonstration of virus particles. Finely cutting, grinding or expressing the diseased material in the staining solution may also yield excellent results. Of course, these methods can only be applied for viruses that remain stable outside their hosts. But with milder stains than phosphotungstate, such as ammonium molybdate and methylamine tungstate several even labile viruses can be made visible. The tests are often more sensitive than serology and sometimes even infectivity tests (Sampson & Taylor, 1968).

The techniques of fixation, embedding, electron-dense staining and ultra-thin sectioning (Milne, 1972) enable the study of viruses *in situ* (reviewed by Matsui & Yamaguchi, 1966; and Esau, 1967, 1968). Much of this work has been done to gain insight into the sites of the virus in the cell and even into the sites of their synthesis (Schlegel *et al.*, 1967). It has also led to the detection of new viruses, in particular the bacilliform viruses, that had escaped notice.

Size and form of plant viruses are constant and therefore characteristic. They are *rod-shaped* and 50–300 nm long and 18 to 25 nm wide (Fig. 45A, B), *thread-like* and more or less flexuous and 480–1250 nm or even about 2000 nm long and 10–11 nm wide (Fig. 45B, C, D), or *spherical or polyhedral* varying in size from 17 to about 80 nm (Fig. 46B, C, D) and in a few cases even *bacilliform* and about 20 to about 60 nm long and 18 or 19 nm wide (Fig. 46A). These viruses consist only of the infectious nucleic acid and a protective protein coat, although evidence is increasing that the big spherical viruses as tomato spotted wilt virus (Fig. 46D) may be more complex in

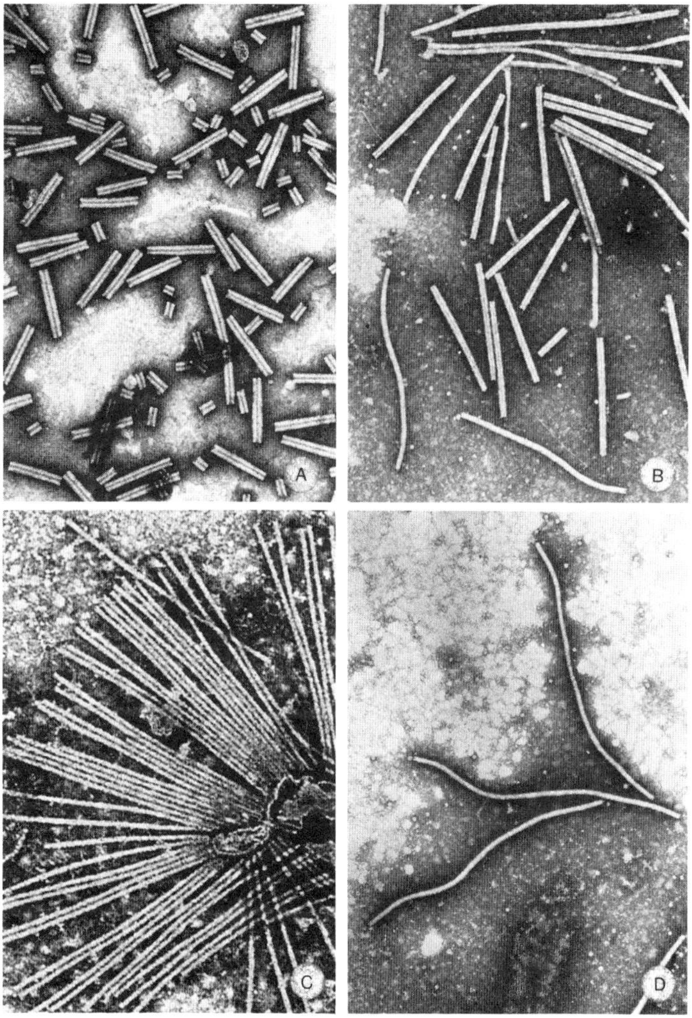

Fig. 45 Electron micrographs of elongated viruses negatively stained with phosphotungstate (× 50 000). A. Tobacco rattle virus consisting of long and short components; B. tobacco mosaic virus (stiff rods) and potato virus X; C. unidentified pea virus from potato virus S group; D. *Wisteria* vein mosaic virus. A. purified preparation, B-D. dip preparations. (All by Technical and Physical Engineering Research Service, Wageningen; A. from Ir H. Huttinga, Wageningen.)

structure, especially in having an extra envelope. This clearly holds for the rapidly enlarging group of relatively big *bacilliform* viruses about 200–300 nm long and 50–100 nm wide (Fig. 57). In negatively stained preparations the particles of these viruses often appear bullet-shaped, but this is thought to be the result of preparative artifacts (Francki, 1973).

All viruses carry the genetic information in the enclosed nucleic acid strand, which is like a chromosome carrying a series of genes. Several viruses, such as tobacco necrosis virus and its satellite, tobacco rattle virus with long and short particles (Fig. 45A), and alfalfa mosaic virus with at least four different components (Fig. 46A), are now known to have their genetic information distributed over two or more independent chromosomes. Here the virion concept, assuming one basic type of particle for each virus, no longer holds (van der Want, 1969). The particles of the more recently discovered viroids, such as of potato spindle tuber, citrus exocortis and chrysanthemum stunt, only consist of naked nucleic acid. They are short RNA strands of ca. 100 000 daltons in weight (Diener, 1972) hardly sufficient to code for one gene. Although these particles can be seen with the electron microscope after purification, they have not yet been observed *in situ*.

Though these details do not directly bear on symptoms of virus diseases, they have to be mentioned, since the virus particles themselves are products of the interaction between plant and virus. Moreover, their presence is characteristic of the host's abnormality, not so much because they can be easily detected by electron microscopy in the sap of diseased plants, but especially because they may constitute a major part of the inclusion bodies. Non-aggregated particles may also be observed *in situ*, although usually with more difficulty.

Intracellular inclusions

Intracellular inclusions, commonly named *inclusion bodies*, caused by virus infection, have been known for a long time. In 1903, Iwanowski discovered two types in the cytoplasm of mosaic-diseased tobacco: *amorphous inclusions* and *crystalline inclusions*. The second type was long neglected because the crystalline material was supposed to be of no etiological significance. The first category attracted more attention because they resembled amoeboid organisms in structure. Their unknown nature later led to the name X bodies, coined by Goldstein (1924).

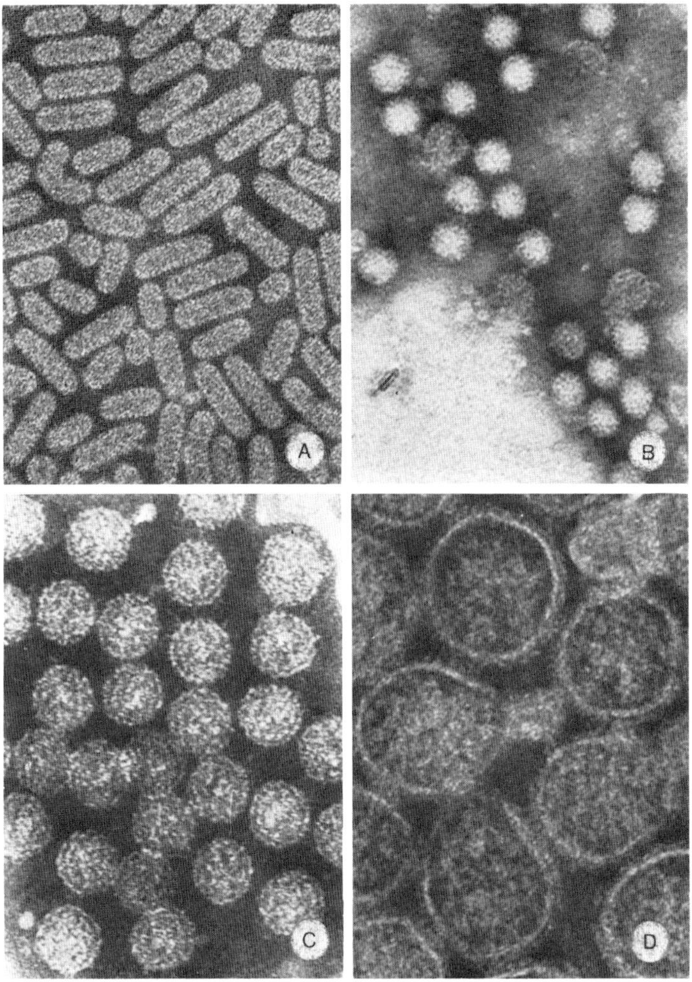

Fig. 46 Electron micrographs of bacilliform and 'spherical' viruses negatively stained with phosphotungstate (× 200 000). A. Alfalfa mosaic virus, B. carnation mottle virus, C. carnation etched ring virus, D. tomato spotted wilt virus. A. purified preparation; B, C. dip preparations. For the enveloped large bacilliform viruses, see Fig. 57. (By Technical and Physical Engineering Research Service, Wageningen. A. From Dr. M. Verhoyen & Mr S. Henstra, Wageningen; D. from Ir T. S. Ie, Wageningen.)

With their isolation and their study by electron microscope (see below), most of them have proved to consist mainly or partly of virus particles. The interest in these structures, formed after infection, is rapidly increasing. So has the number of viruses known to produce inclusion bodies: Sommereyns (1967) listed 45 of them. Information on inclusion bodies has been reviewed by Smith (1958), Goldin (1963), Bawden (1964), McWhorter (1965), Rubio-Huertos (1972), Martelli & Russo (1977) and Christie & Edwardson (1977). The latter authors have reproduced a great number of photographs in colour. Inclusion morphology and structure are usually highly characteristic of the incitant viruses, irrespective of host. Thus they may be of great diagnostic value. Their light microscopy is discussed under Observation, Morphology and distribution, and Development. Much information on their Structure and composition has been revealed by electron microscopy.

Observation Inclusions can be easily observed in epidermal strips from the underside of leaves, but easiest from main veins, petioles or young herbaceous stems, because of larger size and more regular (elongate) shape of such cells (Figs 47, 49, 53) than those of laminar leaf parts, and in leaf hairs (Figs 50A, 51). Internal tissues must be sectioned, e.g. phloem parenchyma and companion cells, for study of beet yellows virus (Esau, 1960a, b) (Fig. 52).

The crystals can often be observed directly in undisturbed living cells in strips or sections mounted in physiological saline under vacuum (McWhorter, 1951). Many crystals in living cells become visible in polarized light, and phase-contrast microscopy has proved especially helpful in examining the development of inclusion bodies inside the living cell (Bald & Solberg, 1961, 1964). Cine-films have been made of cytoplasmic changes leading to the production of inclusions (Sheffield, 1931; Singh & Hildebrand, 1966).

Where these methods fail to distinguish inclusions from normal cell components, several stains have been used. Phloxine staining, introduced by Rubio-Huertos (1950), has proved extremely simple and rapid for fresh tissues, and I had excellent results with a slight modification, for instance by using a mixture of 1% phloxine and 1% methylene blue both in equal volumes of distilled water, ethylene glycol monomethyl ether and ethanol (according to Christie, 1967) for better penetration (Plate 11) (Bos, 1969). For further details on staining techniques see Rubio-Huertos (1972) and Christie & Edwardson (1977). The latter authors mainly apply two procedures. One is with thiazin dyes, such as 0.1 % Azure A in 2-methoxyethanol mixed prior to staining 9:2 with 0.2 M disodium phosphate in distilled water. This procedure is best suited for inclusions with high concentrations of ribosomes and virus particles. The other procedure is with 1 % calcomine orange 2 RS and 1 % 'Luxol' brilliant green BL, both in 2-methoxyethanol, mixed just before staining with water: one part of water, one part orange, and eight parts green. This combination is the most useful for the detection of proteinaceous inclusions. Both procedures are often used in sequence.

Virus inclusions may be confused with normal crystals, as of calcium oxalate, but the latter are easily dissolved in weak acids and do not react with protein stains. The non-virus crystals resist mechanical damage better than virus crystals do.

Morphology and distribution Amorphous and crystalline inclusion bodies have both been found in roots, stems, leaves and flowers of infected plants, and in all tissues except phloem sieve elements and apical meristems. The easiest way is to look for them in epidermal cells or trichomes of leaf and stem, but they are seldom distributed evenly in tissues but rather highly erratic.

Their presence always coincides with virus infection, though not always with externally visible symptoms, as with dahlia mosaic virus in various dahlia varieties (Robb, 1963). As far as they are visible under the light microscope they range in size from about 1 to 30 μm. They differ considerably in shape, position and number per cell, according to the virus. As their presence can easily be demonstrated, they are valuable in diagnosis, as shown by Robb (1963) in indexing dahlias for dahlia mosaic virus.

The amorphous cytoplasmic inclusions often but not always occur near the nucleus, either brought there by converging protoplasmic streams or, as Robb (1964) supposed for dahlia mosaic virus, originating from the nucleus.

They may be granular and often vacuolate, as with bean yellow mosaic virus (Fig. 47, Plate 11A) and several other members of the potyvirus group (Figs 48, 53). Frequently such inclusions are more fibrillar or contain fibrillar or needle-shaped elements (see also p. 135) as in crucifers after infection with turnip mosaic virus (Štefanac & Miličić, 1965), in *Cucurbita pepo* with watermelon mosaic virus

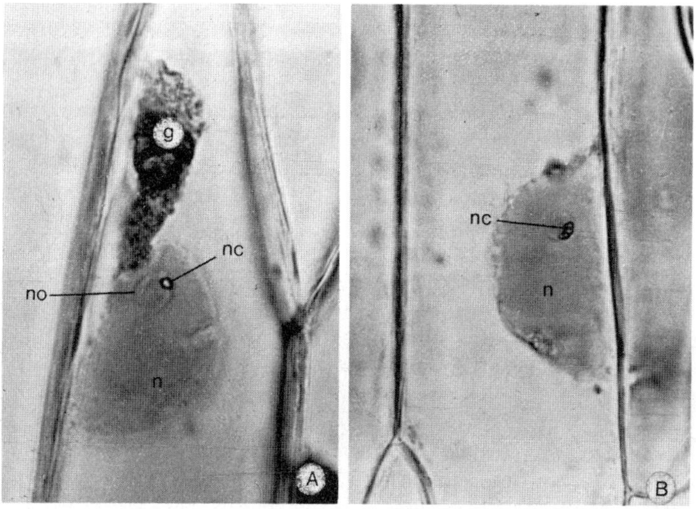

Fig. 47 Granular cytoplasmic inclusions (g) together with tiny nucleolar crystals (nc) stained with phloxine and methylene blue in epidermal cells of broad bean *(Vicia faba)* with a pea strain of bean yellow mosaic virus. n = nucleus; no = nucleolus; (× 500). (After Bos, 1969.)

(Edwardson *et al.*, 1968) or in *Nicotiana clevelandii* and some other plants with sharka (plum pox) virus (van Oosten & van Bakel, 1970, Fig. 48).

Amorphous inclusions may also be round, oval or elongate, but highly refringent and sharply defined from the cytoplasm as with dahlia mosaic virus (Rubio-Huertos, 1950, 1956; Fujisawa *et al.*, 1967) and cauliflower mosaic virus (Plate 11B, D).

In other cases extensive parts of the cells may stain, while the stained areas are not always sharply defined inclusions, as with pea streak virus (Bos & Rubio-Huertos, 1972).

When crystalline, they may be hexagonal and hyaline, as with red clover vein mosaic virus, where there may be several per cell, or they are composite up to 20 μm in diameter and 9 μm thick (McWhorter, 1965; personal observations, Fig. 49). With tobacco mosaic virus, they may be irregular, hexagonal (Fig. 50A) or cubical (Plate 10C), or with Holmes' ribgrass strain (Miličić *et al.*, 1968), and cucumber

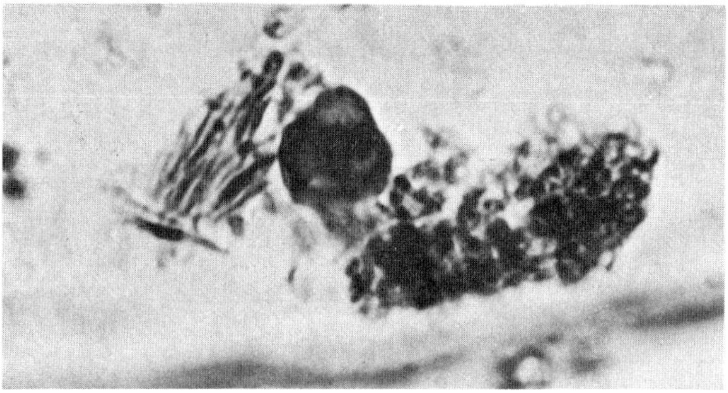

Fig. 48 Granular inclusions and cluster of needles stained with calcomine orange and brilliant green (1:4) in *Nicotiana clevelandii* with sharka (plum pox) virus (× 975). (After van Oosten & van Bakel, in preparation.)

green mottle mosaic virus and *Odontoglossum* ringspot virus (Miličić & Štefanac, 1971) rounded plates. They may be separate, stacked irregularly, or form true hexagonal prisms and become so abundant that they occupy the greater part of the cell. Cells of solanaceous plants infected with tobacco ringspot virus frequently contain many biaxial crystals or rectangular blocks (Bawden & Sheffield, 1939).

Paracrystalline inclusions in cacti may be very large and spindle-shaped (Fig. 50B, C), as already noticed by Molish in 1885. They led to the first discovery of a virus in this family, when Rosenzopf (1951) showed it to be infectious and filterable, and Amelunxen (1958) proved it to consist of a nucleoprotein. For this usually symptomless virus, McWhorter (1965) tentatively proposed the name cactus inclusion virus.

Needle-shaped paracrystals sometimes occur with tobacco mosaic virus. When tobacco cells are treated with dilute acid, TMV crystals disintegrate into numerous paracrystalline needles like those formed in purified suspension by lowering the pH (Beale, 1937). With tobacco mosaic the paracrystalline material may be arranged in fibres many times longer than the cell and curved into figures of 8 (Fig. 51; Kassanis & Sheffield, 1941). Such elongated inclusions, especially when sharply pointed, may hamper cell division and even

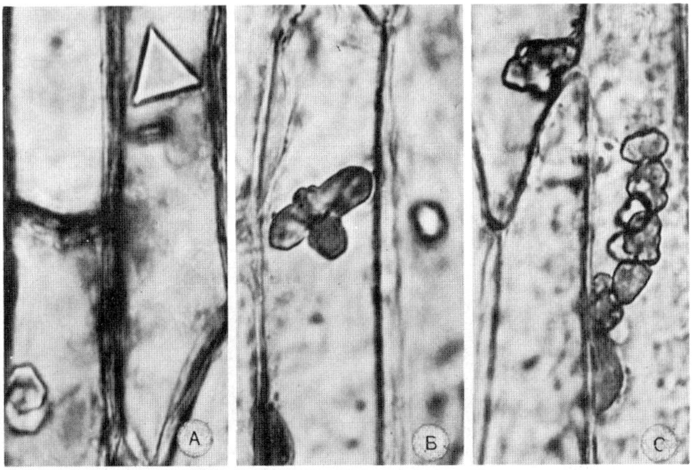

Fig. 49 Crystalline inclusion bodies of various shapes in petiole and stem epidermis of pea *(Pisum sativum)* with red clover vein mosaic virus stained with phloxine and methylene blue; n = nucleus; (× 700). (By Ir R. E. Labruyère, Wageningen.)

fracture membranes and cell walls (reviewed by Goldin, 1966).

The very conspicuous inclusion bodies in sugar-beet infected with beet yellows virus (Esau, 1960a, b) are sometimes homogeneous, in other cases strinkingly banded (Fig. 52), or they consist of aggregates of fibrous material.

With various viruses, only amorphous bodies are produced, whereas with others, there are only crystalline inclusions. Some other diseases are characterized by both, and with bean yellow mosaic virus both are formed as well as the intranuclear crystals (Fig. 47).

It was long thought that crystals occur only in the cytoplasm, but intranuclear inclusions occur too. With tobacco etch virus they are even confined to the nucleoplasm (Kassanis, 1939), where there may be as many as 30 rectangular plates in one nucleus (Sheffield, 1941). With bean yellow mosaic virus and its pea mosaic strain, one or more hexagonal crystals are often found inside the nucleolus (Fig. 47; McWhorter, 1941). Recently I found that various viruses of the potato virus Y group, related to bean yellow mosaic virus, caused a complex of changes in the nucleolus (Bos, 1969); the nucleolus often con-

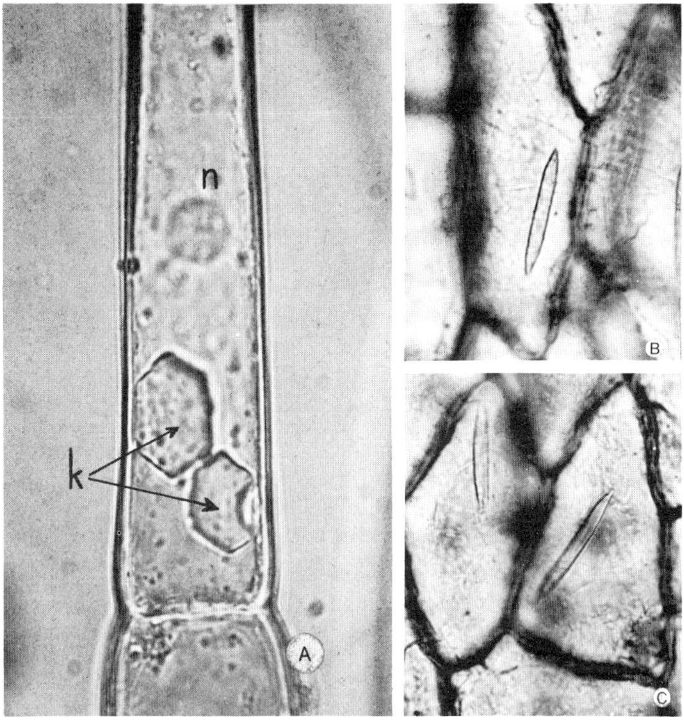

Fig. 50 Unstained preparations of crystalline inclusion bodies of tobacco mosaic virus in tobacco hair cell (A × 350) and of fusiform inclusions presumably of cactus virus X in epidermal cells of Christmas cactus *(Epiphyllum)* (B, C × 275). (A from Lab. Virology, Agric. Univ., Wageningen; B, C by Ir R. E. Labruyère, Wageningen.)

siderably enlarges or sprouts yeast-like, and sometimes internal crystals were present. With a pea necrosis virus, the situation is most unusual: the nucleolus is much enlarged and a mass of crystalline needles protrudes in an odd way from its periphery, or almost the entire nucleus is filled, with some needles pushing out the nuclear membrane (Fig. 53; Bos & Rubio-Huertos, 1969). The latter virus is now considered to be a strain of the clover yellow vein virus. With certain other strains, the enlarged nucleolus may nearly fill the entire

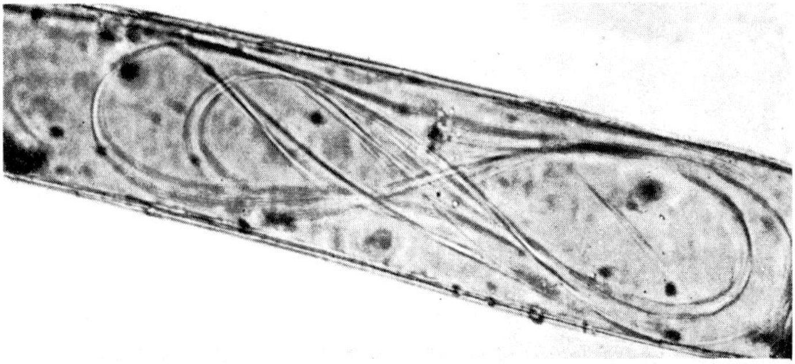

Fig. 51 Long fibrous inclusion bodies curved to figures of eight in halr cell of tomato plant with the tomato aucuba strain of tobacco mosaic virus (× 300). (After Kassanis & Sheffield, 1941.)

nucleus (Bos *et al.*, 1977).

Only recently the occurrence of inclusions in vacuoles has been reported for tobacco mosaic virus (Esau & Cronshaw, 1967), for artichoke mottled crinkle virus (Russo *et al.*, 1968) and for alfalfa mosaic virus (Hull *et al.*, 1969).

Development Inclusions usually develop gradually, though sometimes very rapidly: Zech (1952) has found them 88 hours after mechanical inoculation of tip cells in basal cells of tobacco hairs. With bean yellow mosaic virus in broad beans, Rich (1949) found them one or

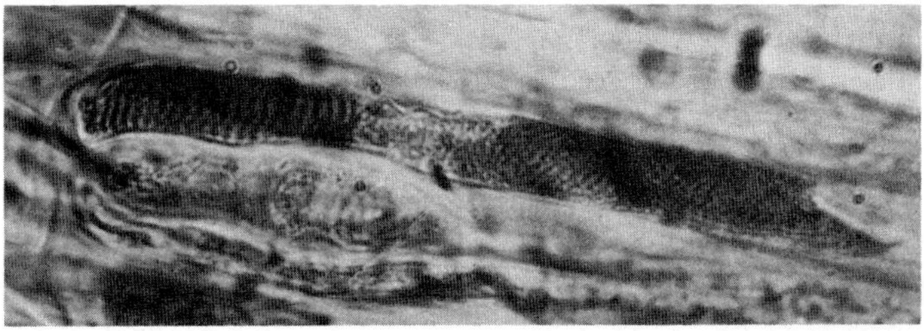

Fig. 52 Cross-banded inclusion bodies stained with Rose Bengal in phloem tissue of beet *(Beta vulgaris)* with beet yellows virus (× 850). (From Dr H. Huttinga, Wageningen.)

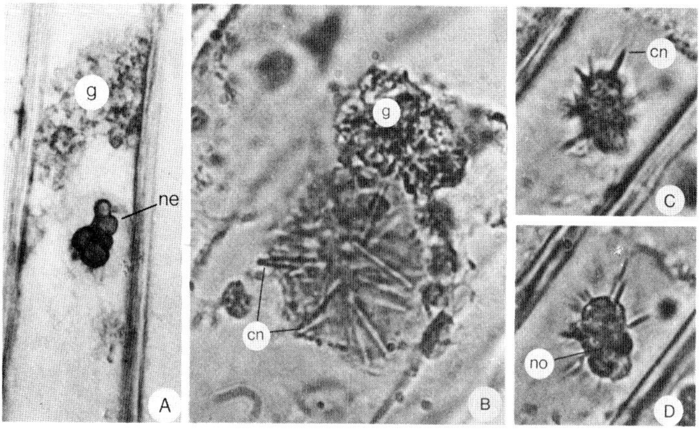

Fig. 53 Granular inclusions (g), nucleolar enlargements (ne) and nuclear crystalline needles (cn) stained with phloxine and methylene blue, and radiating from the enlarged nucleolus (no) of pea *(Pisum sativum)* (A, C, D) and *Nicotiana clevelandii* (B) with pea necrosis virus (A × 235, B-D × 425; C and D at different focus). (Modified after Bos & Rubio-Huertos, 1969.)

two days before the first external symptoms, but usually they do not appear before external symptoms. With severe etch disease in tobacco they appeared two days after external symptoms. (Sheffield, 1941). With sharka (plum pox) virus in inoculated *Nicotiana clevelandii* needle-shaped inclusions were found in the nucleus 10 days after the first external symptoms appeared. In the cytoplasm bundles of needles and granular inclusions arose 14 and 18 days, respectively, after external symptoms became visible (van Oosten & van Bakel, 1970).

Later inclusion bodies may disappear again, as with alfalfa mosaic virus in tobacco (Desjardins, 1966) or amoeboid bodies may become crystalline, as with tobacco mosaic virus in tobacco and tomato (Kassanis & Sheffield, 1941), and in older cells the crystalline structures seem to predominate. These alterations have been attributed to changing conditions inside the cells, especially to progressive production and accumulation of virus particles and their proteins, as will be discussed below. In *N. clevelandii* the intranuclear needles of sharka virus disappeared shortly before or after the first appearance of granular cytoplasmic inclusions (van Oosten & van Bakel, 1970).

Structure and composition Though viruses have always been supposed to participate in the production of inclusion bodies, there have been many speculations about their exact nature. They all react with protein stains, but not with Feulgen's reagent staining deoxyribonucleic acid. Several crystals are presumed to consist of virus particles arranged in a crystalline pattern, but not all react with ribonucleic acid stains. It should be remembered that there are no specific stains for viruses.

Fortunately, much fruitless discussion has been ended by electron microscopy and ultrathin sectioning along with special fixation and staining techniques, allowing direct study of inclusion ultrastructures. These methods have also revealed other accumulations of virus imperceptible by light microscopy. Investigations in many laboratories are now concentrating on this rewarding field of study. For surveys of the rapidly increasing amount of literature see Matsui & Yamaguchi (1966), Schlegel *et al.* (1967), and especially Esau (1967, 1968), who has made fascinating photographs of viruses in their host plants, especially of tobacco mosaic virus and beet yellows virus.

Rubio-Huertos (1950, 1954) and Steere & Williams (1953) first proved that the hexagonal crystals of tobacco mosaic virus consist of virus particles. They removed entire crystals and mounted them on an electron-microscopic grid. A few years earlier, Sheffield (1946) had already extracted and suspended X bodies of tomato aucuba mosaic virus (closely related to tobacco mosaic virus) and shown by electron microscope that they contain many of the characteristic rods, but also other then unidentified material. This was supported by Black *et al.* (1950), who first made ultrathin sections of such X bodies.

It is now known that the 300 nm long particles of tobacco mosaic virus are vertically arranged in layers that may be stacked in 3-dimensional aggregates (Fig. 54; Warmke & Edwardson, 1966). Similar crystals occur in nuclei. Less orderly structures are also known for tobacco mosaic virus. The particles in adjacent layers may be packed at a slight angle. Moreover, the monolayers may aggregate irregularly, thus entrapping other cell components. Rather than side by side, the virus particles may be arranged lengthwise, which results in bundles, rods, needles and fibres, mentioned before.

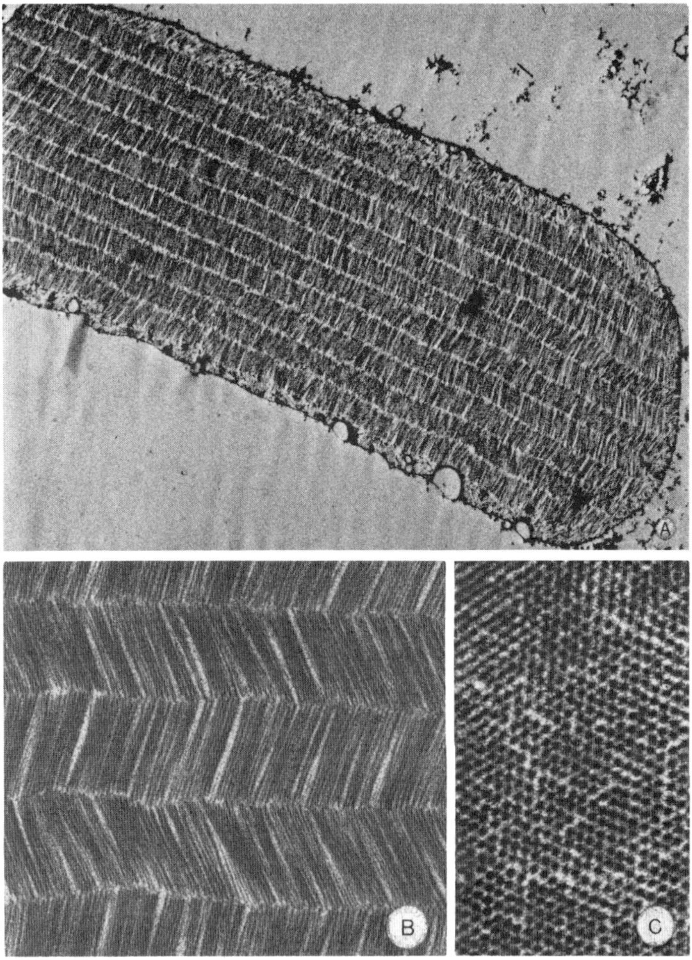

Fig. 54 Ultrathin section (× 11 750) through a hexagonal crystal in leaf parenchyma cell of *Nicotiana tabacum* with tobacco mosaic virus (A). Note the regular arrangement of virus particles in stacked layers. Insets B (× 43 000) and C (× 105 500) show individual particles, C in cross section. (From Dr H. E. Warmke & Dr J. R. Edwardson, Gainesville, Fla; A after Warmke & Edwardson, 1966.)

The banded inclusions of beet yellows virus consist of 1250 nm long particles arranged in layers, though less orderly than those of tobacco mosaic virus because of their greater length and flexibility (Fig. 55; Esau, 1968). The other inclusions associated with beet yellows are a random mass of particles with degrees of orderliness, but not in a regularly repeating pattern (Esau *et al.*, 1966; Esau, 1968).

The striking three-dimensional intracytoplasmic crystals caused by red clover vein-mosaic virus (Fig. 49) were found upon ultrathin sectioning to show a very regular internal striation and to consist of spherical or polyhedral particles 11–12 nm in diameter (Rubio-Huertos & Bos, 1973).

The exact nature of the small crystals sometimes observed in the granular intracytoplasmic inclusions or in the nucleus after infection with certain potyviruses, such as tobacco etch virus (Rubio-Huertos & Hidalgo, 1964) and of bean yellow mosaic virus, is not yet clear. With the latter virus they show an internal periodicity of 7 nm (Weintraub & Ragetli, 1966) and by enzyme digestion they were found to consist entirely or primarily of protein, and no evidence of nucleic acid was obtained (Weintraub & Ragetli, 1968). Bos & Rubio-Huertos (1969) could not detect potyvirus particles in the radiating needles in the nuclei of plants with the pea necrosis strain of clover yellow vein virus (Fig. 53). The crystalline nuclear inclusions induced by tobacco etch virus have been partially purified. They consisted of proteins with molecular weights between ca. 50 000 and 101 000. They were serologically distinct from the coat and pinwheel proteins of the virus as well as from host protein (Knuhtsen *et al.*, 1974).

Amorphous inclusion bodies seem rather complicated. The granular or fibrillar intracytoplasmic inclusions highly characteristic of the potyvirus group have been intensively studied since the mid sixties. Such bodies (Figs 47, 48) contain various so-called cylindrical inclusions consisting of plates, rolls, or curved plates converging to a central core. Sections of these electron-dense structures look like rings, bands, and 'pinwheels' depending on the angle of sectioning (Fig. 56) but they do not consist of aggregated virus particles (Edwardson *et al.*, 1968). With parsnip mosaic virus in coriander the radiating plates forming the pinwheel arms were found to be exceptionally long and flexuous (Murant & Roberts, 1971). The bands and bundles are sometimes visible with the light microscope as

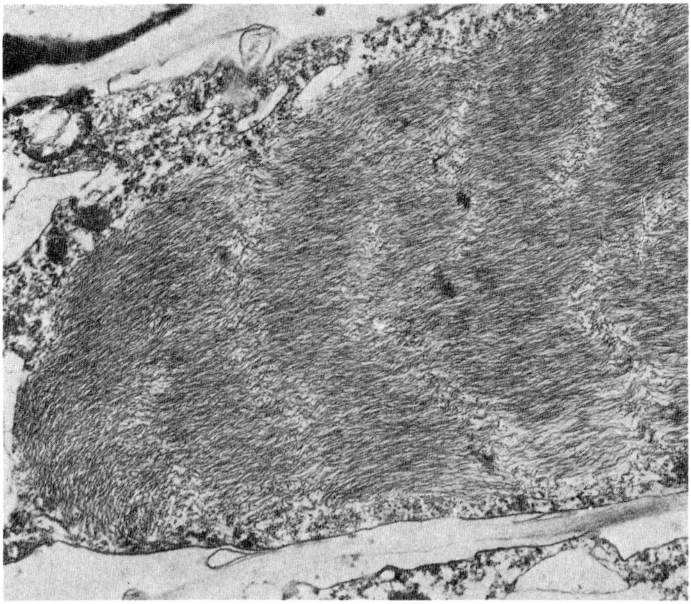

Fig. 55 Ultrathin section (× 18 600) of inclusion in phloem parenchyma cell of beet *(Beta vulgaris)* with beet yellows virus (see also Fig. 52), demonstrating that the bands are layers of long slender particles. (From Dr Katharine Esau, Univ. California, Santa Barbara, Calif.)

fibrils or needles (Fig. 48) and then even pinwheel structures may occasionally be observed (Edwardson *et al.*, 1968). Edwardson (1974), when compiling information on the potyviruses has subdivided them into three groups on the basis of the morphology of the cylindrical inclusion components.

Pinwheel structures of a number of potyviruses have been partially purified from their hosts (Hiebert *et al.*, 1971) and are now known to consist of protein subunits with molecular weights of ca. 70 000 (Hiebert & McDonald, 1973) which is about twice as much as that of coat protein. Detailed serological investigations as summarized by Purcifull & Batchelor (1977) have revealed for instance (1) that the inclusion body proteins are different from viral coat protein and host proteins, (2) that five potyviruses induced inclusion proteins different from those of one another, (3) that such proteins are characteristic of a particular virus irrespective of the host species, and (4) that inclusion proteins of two different viruses (tobacco etch virus end potato virus Y) can both be detected in extracts from double-infected plants. Although the pinwheel components are produced abundantly in potyvirus-infected tissues, their functions are still unknown.

Amorphous inclusions may also contain tubular elements, which have been specially studied for tobacco mosaic virus where they consist of non-infectious protein (e.g. Esau, 1968). This was thought to be identical to the 'X protein', discovered earlier in expressed sap by Takahashi & Ishii (1953). The latter protein is similar in structure to the protein of the tobacco mosaic virus particle. The X tubules are almost twice as wide as normal TMV particles, vary in length, and usually occur in bundles of three. Such tubules have also been found with other viruses, such as a strain of turnip yellow mosaic virus, where they are 110–125 nm wide and have been isolated by ultracentrifugation (Hitchborn & Hills, 1968). The protein was found serologically related to the turnip virus. The tubes are interpreted as 'mistakes' in protein assembly, resulting from lack of association with intact nucleic acid, or as surplus virus protein, though eventual incorporation into virus particles does not seem excluded. The petunia ringspot virus has been found to produce X bodies in *Vicia faba* containing complete virus particles in crystalline array as well as arranged helically in rod-shaped tubules 60 nm wide and more than 700 nm long (Rubio-Huertos, 1968).

The amorphous inclusions may also contain different amounts of virus particles, as already detected by Sheffield (1941) for potyvirus inclusions. Extracted X bodies of tobacco etch virus were found to be infectious.

Masses of virus particles were found in the amorphous inclusions caused by potato virus X, a potexvirus (Shalla & Shepard, 1972). They were found interspersed between alternating laminate inclusion components, consisting of collateral bundles of smooth or beaded sheets which were either slightly curved or rolled into scrolls. The sheets, ca. 3 nm thick, and the beads, ca. 11–14 nm in diameter, were antigenically unrelated to the virus and its protein. Bos & Rubio-Huertos (1972) detected massive amounts of particles of the carlavirus pea streak virus, occurring separately or in several bundles of one particle long, in the cytoplasmic condensations that may be visible

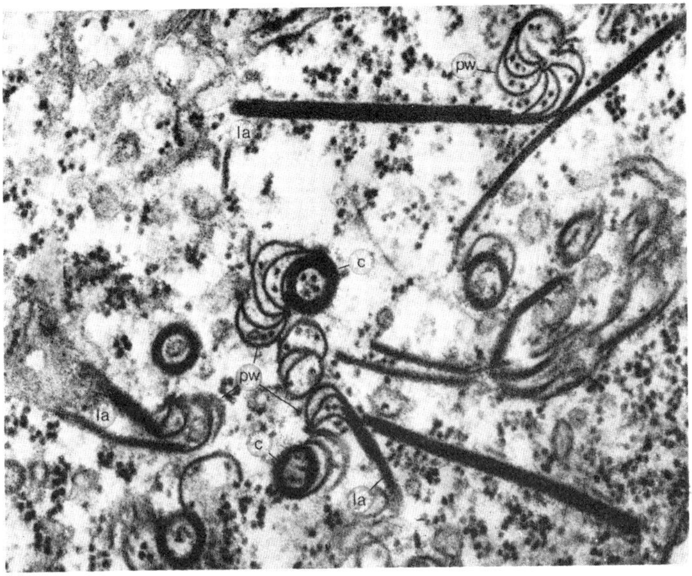

Fig. 56 Ultrathin section (× 60 000) of part of a granular cytoplasmic inclusion in *Brassica perviridis* with turnip mosaic virus; la = laminated aggregate, c = circular inclusion, pw = pinwheel. (After Edwardson & Purcifull, 1970.)

with the light microscope after staining. With their ends these particles were often attached to membranes e.g. around vacuoles.

Warmke (1969) found that the large granular inclusions of the aucuba strain of tobacco mosaic virus in 'Turkish' tobacco plants were made up largely of masses of peculiar angle-layer aggregates and have a high virus content. He found the X bodies (in Goldstein's terminology) produced by common tobacco mosaic virus to be smaller, vacuolate, finely granular and containing relatively few virus particles.

The highly refringent round, oval or elongate intracytoplasmic inclusions by cauliflower mosaic virus have upon ultrathin sectioning been found to consist of many virus particles in a dense, often vacuolate matrix not bounded by a membrane (Fujisawa *et al.*, 1967). Similar structures were found with the related carnation etched ring virus (Rubio-Huertos *et al.*, 1972).

Such clearly defined inclusions have features in common with the so-called *viroplasms* only visible with the electron microscope in cells of plants and leafhoppers infected with wound tumour virus (Shikata & Maramorosch, 1967) and in tumour cells of maize leaves with maize rough dwarf virus (Gerola & Bassi, 1966), where they contain large amounts of virus particles. McWhorter (1965) had previously used the term *viroplasts* for X bodies in general.

In situ, elongate plant viruses often tend to attach by their ends to membranes, as just mentioned for pea streak virus. With a special strain of tobacco rattle virus such an association was found between its rod-shaped particles and mitochondria. The particles are often densely packed side by side around and between abnormal mitochondria, clumping them together into X bodies (Harrison & Roberts, 1968; Harrison *et al.*, 1970).

Other components of amorphous bodies are organelles normally involved in protein synthesis, such as the endoplasmic reticulum and ribosomes, as well as degenerated protoplasm constituents. So the amorphous inclusions seem to be regions of virus-directed protein synthesis, including polymerization of subunits into tubules (Esau, 1968), manufacturing centres and 'storehouses' for viruses in the cell (McWhorter, 1965), virus assembly sites (Shikata & Maramorosch, 1967) or virus factories (Kamei *et al.*, 1969).

Submicroscopic virus aggregates and non-aggregated particles

Several plant viruses are now known to produce an abundance of virus particles, even when no inclusions or accumulations have been detected by light microscopy, or to do so in organelles where they cannot be seen by light microscopy. They are easier to detect when occurring in crystalline array as with rice dwarf virus (Fukushi *et al.*, 1962), artichoke mottle crinkle virus (Russo *et al.*, 1968), and bean southern mosaic virus (Weintraub & Ragetli, 1970).

A number of viruses have been found in nuclei, such as beet yellows virus (Cronshaw *et al.*, 1966) pea enation mosaic virus (also in its vector; Shikata & Maramorosch, 1966; Shikata *et al.*, 1966), tobacco mosaic virus (Esau & Cronshaw, 1967) and tomato bushy stunt virus (Russo & Martelli, 1972). So far, reports on the incidence of virus aggregates in chloroplasts are limited, such as of clover yellow

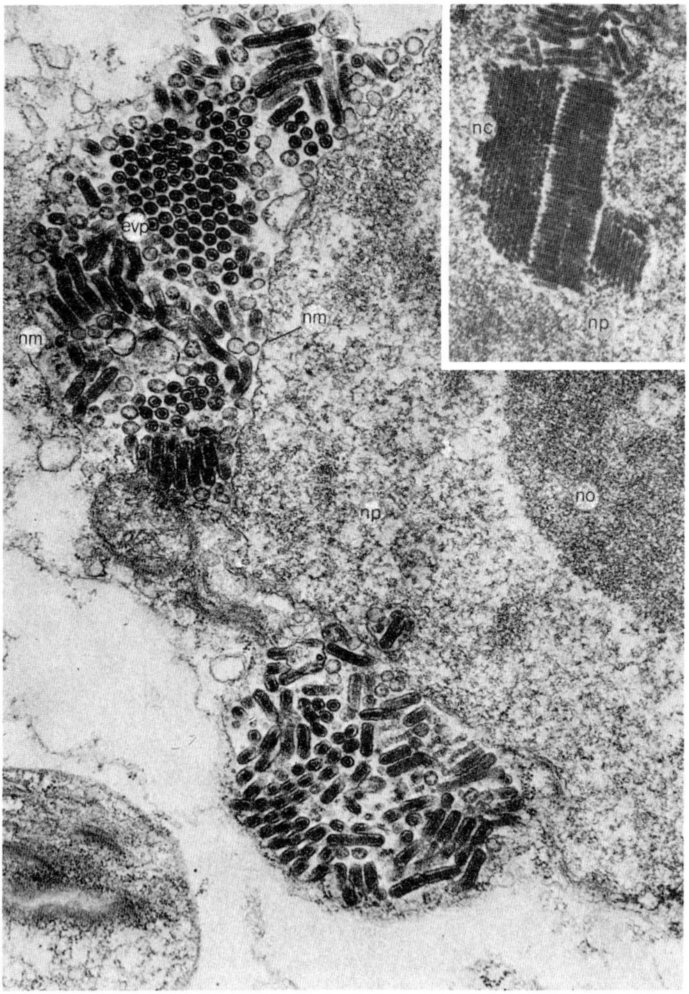

Fig. 57 Accumulations of enveloped bacilliform virus particles (evp) in perinuclear spaces between the nuclear membranes and, in inset, naked nucleocapsids (nc) of the virus in the nucleoplasm (np) of enation-diseased white clover plants (see Fig. 32) (× 20 900); nm = nuclear membranes, no = nucleolus. (Modified from Rubio-Huertos & Bos, 1969.)

mosaic virus in *Pisum sativum* (Purcifull *et al.*, 1966) and of tobacco mosaic virus in tobacco (Esau & Cronshaw, 1967; Shalla, 1968b). Such accumulations may be regular crystals, as of wound tumour virus in the fat body of its vector (Shikata & Maramorosch, 1965). Beautifully layered aggregates of potato virus X particles have even been found in apical meristems of *Datura* plants (Pennazio & Appiano, 1975).

If they occur inside the cytoplasm, small groups of virus particles are often surrounded by sac-like or tubular envelopes, which might partly be derived from the endoplasmic reticulum. Such tubules containing single rows of spherical particles were found in meristematic tissues infected with three nematode-borne polyhedral viruses (Walkey & Webb, 1970). They often occur in inclusion bodies, as with strawberry latent ringspot virus (Roberts & Harrison, 1970), where each tubule, or occasionally two or three, was enclosed in a membraneous sheath joined to the endoplasmic reticulum. Similar tubules were also observed in viroplasms, as with maize rough dwarf virus (Gerola & Bassi, 1966). Their nature and function is still obscure.

Recently the big bacilliform or bullet-shaped viruses, often called rhabdoviruses, have attracted special attention, e.g. lettuce necrotic yellows virus, potato yellow dwarf virus, wheat striate mosaic virus, sowthistle yellow vein virus, broccoli necrotic yellows virus, rice transitory yellows virus, and a virus isolated from enation-diseased white clover (Fig. 57; Francki, 1973). Some have been found in the cytoplasm only, and then apparently in vesicles of the endoplasmic reticulum, some only in the nucleoplasm and in perinuclear spaces, whereas others have been detected in both cytoplasm and nucleus.

The bacilliform viruses consist of an inner core, the nucleoprotein (nucleo-capsid), and an envelope. Naked inner cores are found only in the nucleoplasm and then often in crystalline array. They are supposed to be coated while passing the inner nuclear membrane when entering the perinuclear space between the two lamellae of the nuclear membrane (e.g. Kitajima & Costa, 1966; MacLeod *et al.*, 1966).

With the electron microscope non-aggregated particles may be made visible too *in situ*, although recognition of elongated viruses is often hard when they are cut obliquely. Such single particles are most frequently found in amorphous inclusions, but also randomly distributed in the cytoplasm and sometimes in the vacuole. They are often associated with the cisternae of the endoplasmic reticulum, e.g. the spherical particles of the Brazilian strain of the tomato spotted wilt virus (Kitajima, 1965). Bundles of the long and flexuous particles of beet yellows virus have even been observed in plasmodesmata between parenchyma cells, thus demonstrating that intact particles may pass unchanged from cell to cell. In sieve elements they may fill the sieve-plate pores (Esau *et al.*, 1967). Likewise elongate particles of potato virus Y, tobacco etch virus and tobacco mosaic virus (Weintraub *et al.*, 1976) and spherical particles as of chrysanthemum aspermy virus (Lawson & Hearon, 1970) were detected in plasmodesmata of their hosts. Small isometric viruses if occurring separately are easily confused with normal isometric plant constituents.

Thus, much information on the distribution of the viruses in hosts is already accumulating. Occurrence or accumulation at a particular site, however, does not prove synthesis or polymerisation there. With bacilliform viruses the occurrence of simple forms inside the nucleoplasm and of 'mature' particles in perinuclear spaces and cytoplasm strongly suggests their nuclear origin. However, the situation may differ from virus to virus, and the site of virus synthesis can also be investigated by other techniques, like ultraviolet micro-spectrophotometry and immuno-microscopy, either using fluorescent antibodies for light microscopy or electron-dense ferritin or heavy metals or radioactive iodine for electron microscopy. A discussion of these still provisional techniques is beyond the scope of this book, and 'conclusive proof that virus synthesis takes place at a particular site has not been possible with any of these methods' (Schlegel *et al.*, 1967). These techniques are mentioned here to demonstrate how technological progress has allowed symptomatology to expand rapidly to include observations on the viruses themselves, almost at the molecular level. These deviations are now perceptible to man; thus strictly spoken they are symptoms of infection.

Search for virus particles in ultrathin sections of vascular tissue has revealed no such virus particles but masses of mycoplasmas (Fig. 58). They were first detected by Doi *et al.* (1967) in Japan. Like viruses, they have been seen massing at sieve-plate pores and hence seem to be passively transported in vascular tissue. The number of plant diseases, mainly leafhopper-borne, now being associated with these pleomorphic, mostly submicroscopic organisms is constantly

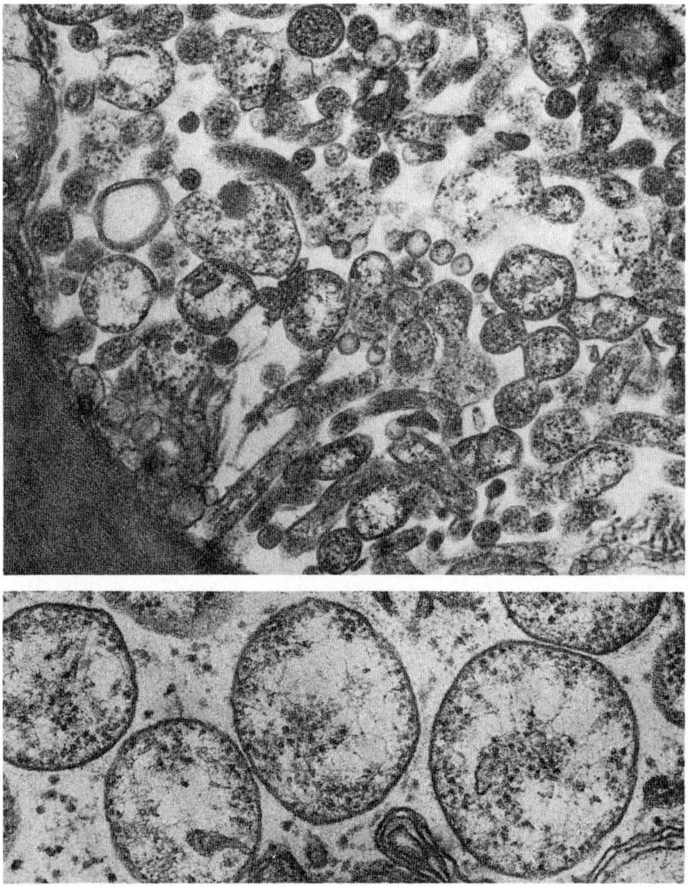

Fig. 58 Above: abundant mycoplasmas ranging in diameter from 80–800 nm in part of a phloem cell of wild tobacco *(Nicotiana rustica)* with aster yellows (× 36 500). Below: some of the largest bodies (× 50 000) showing bounding membrane, ribosomes and strands of nucleic acid. (From Dr H. Hirumi and Dr K. Maramorosch, Boyce Thompson Institute for Plant Research, Yonkers, N.Y.)

increasing and has already passed 50 (p. 105). They have also been detected with certain diseases mainly characterized by phloem degeneration and ensuing atypical growth reduction, discoloration, wilting and premature death, such as pear decline (psylla-transmitted) and elm phloem necrosis.

Recently, the mycoplasma-like organisms of corn stunt (Davis & Worley, 1973) and citrus stubborn (Cole *et al.*, 1973) were demonstrated to be motile and of spiral shape *(spiroplasma)*. Electron microscopy of some other diseases of suspected virus etiology has also shown the involvement of *Rickettsialike organisms*. These differ from mycoplasmas in having a rigid cell wall and being round or elongated in cross section. They were, for example, found in phloem of clover with club leaf (Windsor & Black, 1972) and in xylem of grapevine with Pierce's disease and lucerne with alfalfa dwarf (Goheen *et al.*, 1973).

Deviations resembling symptoms of virus diseases

As the reader should by now be aware, certain virus symptoms closely resemble deviations due to other causes. This needs some further elaboration to avoid confusion when visually diagnosing diseases.

Genetic aberrations

Several types of variegation attracted attention long before the discovery of viruses. Their presumed genetic character was reflected by names like *Abutilon striatum* 'Thompsonii' for mosaic-diseased plants (Fig. 2), and *Lonicera japonica* 'Aureo-reticulata' for plants with a bright vein mosaic or irregular vein yellowing (e.g. Schmelzer & Schmidt, 1963a, 1963b). Both species are easily propagated vegetatively and only the infected 'form' is of commercial value. Such deviations have long been considered mutants. Even though some of them were found to be transmitted by grafting their true nature was not understood (e.g. a paper by Masters (1877) on 'Action of scion on stock') until the virus concept was defined (Baur, 1904).

Many virus-induced abnormalities closely resemble genetic deviations because viruses act like genes in influencing the host via their nucleic acids. On this resemblance Duggar & Armstrong (1923) have even proposed a gene theory of the nature of viruses, for which there are still grounds (Atanasoff, 1963).

Some viruses are transmitted by pollen or seed ('pseudo heredity') and several, if not all, forms of cytoplasmic heredity might be attributed to viruses (Atanasoff, 1964).

Sometimes, notably with *Abutilon striatum* 'Thompsonii', plant and virus seem to live in harmony. Some bacterial and animal viruses even enter a non-infectious 'occult' stage by being incorporated into the host's genetic system. Under certain conditions, e.g. of stress, such viruses can escape normal genetic control by the host and resume pathogenic character by 'explosive' multiplication.

To distinguish viral and genetic variegation, Schmelzer & Schmidt (1963b) have pointed out that genetic mosaics usually appear as sectorial or periclinal chimaeras. In the chimaeras certain tissues are free of chlorophyll which, in leaves, gives rise to white, yellow, or chlorotic leaf edges, central parts, or more irregular patterns with sharp borderlines. However, diffuse spots may be either genetic (*Aucuba japonica* 'Variegata') or viral. Line and ring patterns are always caused by a virus, but a permanent vein yellowing throughout a plant is genetic in origin. With viruses the deviations may temporarily disappear, e.g. many types of vein yellowing and vein chlorosis, and are often irregularly distributed such as in case of vein mosaic, whereas genetic deviations have a constant nature.

In young seedlings of some plants such as peas and broad beans, some sort of transitory variegation is sometimes observed, suggesting infection by a seed-borne virus. Later the plants develop normally and I have never been able to isolate a virus from such plants. This phenomenon may be related to 'June yellows' in strawberry, only appearing in spring and autumn during cool weather; it has been artificially induced by coolness and bright light (Braak, 1955). A non-infectious variegation in *Nicotiana glutinosa*, closely resembling virus infection, can be induced by heat (Benda, 1962; John & Weintraub, 1966). The strawberry disorder was later ascribed to a plasmagene whose concentrations depended on external conditions and on genotype of the host. It was thought to be a mutant cell constituent or more probably, an exogeneous virus-like particle intimately associated with cell nuclei, although not transmitted by grafting or insects (Wills, 1962).

Flower colour breaking in tulip may also be genetic in origin and the difference from virus-induced breaking often cannot be determined visually (van Slogteren & de Bruyn Ouboter, 1941).

Plants with leaf narrowing indistinguishable from certain virus diseases (Fig. 37) have often been described as cultivars or formae named *laciniata* or *dissecta*, e.g. *Aesculus hippocastanum*, *Castanea sativa*, *Tilia platyphyllos* and *Quercus petraea*. Especially in *Castanea* and *Tilia*, I have found such malformations to be irregular, also in their distribution over the tree, and often associated with mottling. In tomatoes (Edwardson & Corbett, 1962) and cylamen (Scholten, 1966) some cases of leaf narrowing proved to be genetic and controlled by a single recessive factor.

A similar confusion may arise even with necrosis. In Wisconsin certain inbred lines of cabbage had necrotic flecks or spots on leaves, later tissue of the head broke down, or the base of petioles became severely necrotic, resulting in defoliation. Repeated efforts to demonstrate a virus or to isolate fungi or bacteria failed. The apparently 'autogenous necrosis' was found to be completely recessive and controlled by several genes (Pound & Walker, 1953). Likewise a disorder developing during storage of certain lines of red cabbages in the Netherlands consisted of groups of small necrotic spots, especially along the main veins. It resembled necrotic stipple in white storage cabbage due to cauliflower mosaic virus. It was not caused by virus (van Hoof, 1954) or other pathogens, but was controlled by a single incompletely recessive genetic factor (Nieuwhof & Wiering, 1962). Another genetic disorder of white cabbage showing up during storage is 'grey speck' disease. Symptoms consist of superficial minute grey-black specks on the basal parts of the outer head leaves, or sometimes covering the entire leaf, but mostly along the main veins (Nieuwhof *et al.*, 1974).

Physiological non-parasitic disturbances

In fact, heat-induced types of variegation, and necrosis developing in cabbages during storage just-mentioned, though genetically controlled, are physiological non-parasitic disturbances. *Gomphrena globosa*, which is extensively used as a local lesion indicator host for several viruses, is known to produce spontaneous lesions under high light intensities (Francki, 1967).

Especially difficult to distinguish from deficiency diseases are virus diseases characterized by chlorosis or reddening. Oat red leaf, caused by barley yellow dwarf virus (Rademacher & Schwarz, 1958) can only be distinguished from nitrogen deficiency by its late occurrence in the season. Symptoms of cotton anthocyanosis virus infections resemble those due to magnesium deficiency, and actually magnesium content is lower in diseased cotton leaves (Costa, 1956). In fact, these and other yellowing diseases have long been ascribed to faulty nutrition, poor root development or unfavourable conditions and it is only since recently that a virus like beet western yellows virus has been found to be widely distributed in various crops and wild species in many countries (Duffus, 1977).

The similarity between the effects of the viruses concerned and mere physiological disorders is not fortuitous; many physiological changes caused by virus diseases can be explained as deficiency of essential building blocks, or as premature senescence. It has been shown earlier, how decline diseases are mainly characterized by deficiencies. Lack of carbohydrates caused root deterioration, resulting in decline in function of aerial parts because of lack of water and mineral nutrients. Thus, citrus tristeza and pear decline may resemble symptoms produced by girdling, 'wet feet', malnutrition, winter injury and drought in pear-trees (Batjer & Schneider, 1960).

The effect of herbicides, which contain growth hormone leading to leaf malformation and vein clearing, may be confused with virus infection. Gold & Faccioli (1968) found that various symptoms of beet curly top in several species can be reproduced by administering ethylene to the roots. Traces of residues from the herbicide sodium chlorate in soil growing grapevine, lettuce, cucumber, tomato or cauliflower caused vein chlorosis and irregular vein banding highly reminiscent of virus infection (den Boer, 1958).

Insect toxaemias

Shalla *et al.* (1963) showed that pear decline was not, as previously supposed, caused by the pear psylla *(Psylla pyricola)* as it can be transmitted by grafting. The psylla only acts as a vector, of what is now known to be a mycoplasma, but the induction of slow decline symptoms in pear, especially the reddening of foliage, by pear psylla infestations in the absence of virus could not be excluded (Jensen *et al.*, 1964). This direct toxic effect of psylla feeding was later proved by Griggs *et al.* (1968).

Top-roll in potatoes, characterized by upward rolling round the midrib of leaflets of the upper and occasionally the middle leaves of plants (Fig. 59), can reduce tuber yield by 40%. It greatly resembles a virus disease but was found neither to be graft-transmissible nor tuber-perpetuated and not to be caused by potato leafroll virus. It was only recently demonstrated to be due to intensive feeding by the aphid *Macrosiphum euphorbiae*. Where the aphids had fed, the plant cells were often found to be killed and surrounding cells to be stimulated to multiply. It takes several days after feeding before top-roll symptoms appear. Foliage produced after aphicide treatment was symptomless even if it arose from the axil of an affected leaf (Gibson, 1974).

These examples demonstrate the sort of difficulties encountered in determining the cause of a disease, in distinguishing between virus infection and insect tox(a)emia. When insects or mites are involved

Fig. 59 'Bintje' potato plant with top-roll caused by toxic saliva substances of *Macrosiphum euphorbiae*.

the question is: is their saliva the real origin of the disease when brought into a plant or does it contain a virus as the incitant? For a general review, see Carter (1973); for Homoptera see Nuorteva (1962), for free-living gall mites, see Proeseler (1968).

Besides mechanical injury toxic effects may be either local or systemic, though the translocation of the toxins is only limited. So symptoms may appear in leaves unfolding after the insects have been removed, but the leaves unfolding later are usually normal, unless the affection consists of less easily reversible changes as in mealybug wilt of pineapple (Carter, 1973).

The toxic effect may lead to chlorosis, vein banding, vein chlorosis, chlorotic streaks and stripes, and even to wilting or malformations, among which gall formation is of special importance.

Several toxaemias are indistinguishable from deviations resulting from viruses, as with vein clearing occasionally caused by the beet leafhopper and the curly top virus (Bennett, 1952). Other affections closely resembling virus diseases, but due to free-living gall-mites are chlorotic flecks in *Lolium multiflorum* and *L. perenne* caused by *Phytocoptes hystrix*, asteroid spots in *Prunus* spp. caused by *Vasates fockeui*, and raspberry mosaic-like symptoms in raspberry caused by

Eriophyes gracilis (reviewed by Proeseler, 1968).

Bos & Grancini have (1968) described how the symptoms of clover enation and some related virus diseases closely resemble vein enlargements and vein enations caused by toxic substances of leafhopper saliva. Among these insect toxaemias are the clover enation disease, erroneously described as a virus disease by Maramorosch (1953), an 'ephemeral disease' of maize (Maramorosch, 1959) and 'pseudo Fiji disease' of sugar-cane in Madagascar (Antoine, 1959; Baudin, 1960). In pseudo Fiji disease the absence of phloem proliferation may distinguish it from virus infection. Unlike virus-diseased plants, the specimens affected by insect toxaemias produce completely normal organs a while after removal of the insects.

Disorders due to pathogenic micro-organisms

Sometimes virus diseases have been confused with disorders due to pathogenic micro-organisms, especially where the syndromes result from the combined effects of virus and fungus or bacteria (though the appearance of fungi and bacteria is usually secondary). Reference has already been made to the combination top yellows and *Fusarium solani* in peas, and to stolbur and *Colletotrichum atramentarium* in potatoes. Certain fungi may produce, on their own, syndromes similar to those of viruses, e.g. leafrolling, chlorosis, and anthocyanin formation in potato caused by potato leafroll virus or *Rhizoctonia solani* (Bode, 1968).

Gum secretion on trunks and twigs of apricot trees does not occur only with a certain virus-infection but may also be produced by the bacterium *Pseudomonas syringae*, by the fungus *Phytophthora syringae*, and probably by other pathogens, or even by winter injury and certain insect infestations (Pine, 1963).

The production of local lesions after sap inoculation can also lead to misinterpretation, since in some test plants certain bacteria can cause them. Smith (1951) has described a 'beet latent virus' based on the observation that beet sap may induce necrotic local lesions in the cotyledons of *Vigna sinensis*, but Yarwood *et al.* (1961) in their attempts to characterize this hypothetical virus found the lesions to be caused by the bacterium *Pseudomonas aptata*. The same bacterium is responsible for similar lesions in cowpea after inoculation from mosaic-diseased sunflower plants, and in cowpea, some other legumes and cucumber after back-inoculation from cowpea (Schmelzer &

Fig. 60 Local lesions with a translucent halo on *Chenopodium quinoa* five days after inoculation with sap from lilac containing *Pseudomonas syringae*. (From F. A. van der Meer, Wageningen).

Molnár, 1964). Since then there have been a number of reports (e.g. Opel *et al.*, 1974) on the production of local lesions by *Pseudomonas syringae* on *Chenopodium quinoa* and some other test plants when making inoculations from 'virus-infected' plants (Fig. 60). In my own work with French beans it has sometimes been difficult to distinguish systemic chlorosis and malformation due to *Pseudomonas phaseolicola* from virus mosaics.

Usually deviations caused by micro-organisms are accompanied by accumulations of these organisms or their fruiting bodies. But sometimes this criterion fails because the toxic effects are systemic or restricted to local mechanical damage of vascular tissue with a systemic effect. This is so with several fungi, like *Verticillium albo-atrum* and *Fusarium* spp., where local infection blocks vascular tissues, leading to wilting and yellowing, sometimes resembling a virus infection. If so, a further check on lower plant parts may reveal the parasite.

Multicausal abnormalities

Sometimes different agents lead to identical deviations, especially with malformations. So enations and tumours may result from attacks by bacteria (such as crown gall caused by *Bacterium tumefaciens*) and various mites or insects. They are then called galls. As already mentioned several times, viruses may cause similar structures. Such deviations may also originate spontaneously, as in certain interspecific *Nicotiana* hybrids, especially *N. glauca* × *N. langsdorffii* (Kehr & Smith, 1954).

Presumably most multifarious are the causes of witches' broom phenomena: they may be due to bacteria, fungi, mites, insects, and even to parasitic higher plants such as dwarf mistletoes (*Arceuthobium* spp.) all causing local witches' broom growth, sterility and other floral abnormalities, or to 'viruses' resulting in entirely abnormal plants as described before in detail (Bos, 1957a, 1963). There are indications that some witches' brooms even arise by bud mutation and are thus genetic. This is supported by the wide variation in growth habit of many plant species including fastigiate forms and forms with extremely dense branching or bushy appearance, as in *Viburnum fragans* 'Nanum', and in *V. opulus* 'Nanum' which is even completely sterile (Bos, 1963).

Several systemic witches' broom diseases, formerly ascribed to viruses, have now been associated with mycoplasma-like micro-organisms. Most peculiar, in tomatoes the anatomic effects of aster yellows, one of these diseases, are fundamentally equal to those caused by beet curly top virus (Rasa & Esau, 1961). Likewise, the effects of phloem degeneration caused either by virus (citrus tristeza) or by mycoplasma (pear decline) may be similar, as described on p. 26. There, comparable effects have even been ascribed to the *Psylla* which acts as a vector of pear decline.

Retrospect

The variety of changes in plants incited by viruses is amazing. No species of plant, no part, no function seems to escape their influence.

The effects start at disease onset, the start of the virus-host relationship, leading to changes in host biochemistry, physiology and anatomy, and finally to macroscopically visible deviations. It is hard to classify the abnormalities; even the limits of disease and normality are indistinct.

A further difficulty is that important symptoms like growth reduction, colour deviations, death of tissues, cork formation, and malformation are not specific to virus infection. Each may have other causes, so that virus symptoms are often confused with those from other pathogens and non-parasitic causes.

Physiologically, biochemically and even anatomically the deviations due to viruses do not differ from those associated with degeneration in normal ageing or in deficiencies. This is easily understood, for viruses act through the host's metabolism. They behave like an uncontrolled part of the host's genetic system, deranging normal synthesis of nucleic acid and amino acid by inducing the host to produce almost 'explosively' a mass of virus nucleoprotein. Fortunately, many cases are known where plants can 'fight back', can, to a certain extent, regain control and restore physiological equilibrium by gradually reducing virus multiplication to less harmful proportions (recovery), though they rarely seem able to expel the alien genetic material completely. In a few cases some sort of harmless coexistence results.

Thus, derangement of host metabolism is gradual only, but forms a cumulative sequence so that the final effect may be massive.

Virus diseases emerge in general picture as a highly complex group of symptoms, resembling those of other disorders. However, virus symptoms often have a characteristic distribution in the host and a specific relation to certain tissues, according to the manner of transmission of viruses from plant to plant, to their site and time of introduction, and to their manner and rate of multiplication and internal transport. Symptoms of diseases caused by mycoplasma-like organisms closely resemble those of phloem-limited viruses with leafhopper and aphid vectors in epidemiology and distribution in the plant.

The detailed descriptions of symptoms should help readers in distinguishing virus diseases from abnormalities caused by other influences. They should also remind the reader not to rely solely on symptoms when diagnosing a disease, although standardization of conditions may improve reliability of routine indexing with test plants.

Neither can viruses be distinguished by infectivity, as concluded from reproduction of symptoms after artificial transmission, as recently demonstrated by mycoplasmas transmissible by grafting and by insects. Association with certain virus particles is equally inadequate evidence, even when accumulations can be observed by electron microscope. Complex infections with latent viruses are common, and the virus observed may be the latent component.

Full proof that a given disease, syndrome or symptom is caused by virus can be obtained only by applying Koch's postulates, which have been neglected by virologists because viruses could not be handled or grown on artificial cell-free media (e.g. Rivers, 1937; Schmelzer & Molnár, 1964; Bos, 1965b). Viruses should be isolated, and physicochemically identified outside the natural host and then returned to a healthy, genetically similar host-plant for artificial reproduction of the original effect. In all such research, knowledge of symptoms remains indispensable.

References

Allington, W. B., 1946. Bud blight of soybean caused by the tobacco ring spot virus. – Phytopathology 36: 319–322.
Amelunxen, F., 1958. Die Virus-Eiweissspindeln der Kakteen. Darstellung, elektronenmikroskopische und biochemische Analyse des Virus. – Protoplasma 49: 140–178.
Antoine, R., 1959. Cane diseases. – A. Rep. Mauritius Sugar Ind. Res. Inst.: 53–60.
Arenz, B. & W. Hunnius, 1963. Erfahrungen mit dem Igel-Lange-Test in der Serienarbeit. – Bayer. landw. Jb. 40: 122–136.
Atanasoff, D., 1922. A study into the literature on stipple-streak and related diseases of potatoes. – Meded. LandbHoogesch. Wageningen 26: 1–52.
Atanasoff, D., 1926. The stipple-streak disease of potato. A complex problem. – Izv. bŭlg. bot. Druzh. 1: 43–52.
Atanasoff, D., 1963. Genes and viruses. – Z. PflZücht. 49: 205–209.
Atanasoff, D., 1964. Viruses and cytoplasmic heredity. – Z. PflZücht. 51: 197–214.
Atanasoff, D., 1971. The viral nature of cytoplasmic male sterility in plants. – Phytopath. Z. 70: 306–322.
Atkinson, P. H. & R. E. F. Matthews, 1967. Distribution of tobacco mosaic virus in systemically infected tobacco leaves. – Virology 32: 171–173.
Attafuah, A., 1965. Embryo abortion in virus-infected sweet cherry. – Pl. Prot. Bull. F.A.O. 13: 79–82.
Aubert, O., 1960. Les viroses du tabac en Suisse. – Mém. Soc. vaud. Sci. nat. 12: 153–211.
Baerecke, M. L., 1955. Der Nachweis der Blattrollinfektion bei Kartoffeln durch ein neues Färbeverfahren. – Züchter 25: 309–313.
Baker, E. A. & A. I. Campbell, 1967. Effect of viruses on the pigment composition of apple bark and pear leaf. – A. Rep. Long Ashton Agric. hort. Res. Stn 1966: 141–144.
Bald, J. G. & P. A. Solberg, 1961. Apparent release of tobacco mosaic virus in living infected cells. – Nature (Lond.) 190: 651–652.
Bald, J. G. & P. A. Solberg, 1964. Cytological reactions of normal and TMV-infected tobacco leaf cells to acid and alkaline solutions. – Am. J. Bot. 51: 396–404.
Banttari, E. E. & M. B. Moore, 1962. Virus cause of blue dwarf of oats and its transmission to barley and flax. – Phytopathology 52: 897–902.

Batjer, L. P. & H. Schneider, 1960. Relation of pear decline to rootstocks and sieve-tube necrosis. – Proc. Am. Soc. hort. Sci. 76: 85–97.
Baudin, P., 1960. Observations sur les nouvelles galles foliaires de la canne à sucre à Madagascar. – Revue agric. sucr. Île Maurice 39: 220–228.
Baur, E., 1904. Zur Aetiologie der infektiösen Panaschierung. – Ber. dt. bot. Ges. 22: 453–460.
Bawden, F. C., 1958. Gradations and transitions between pathogenicity and commensalism in infections with plant viruses. – Proc. Symp. on Latency and Masking in viral and Rickettsial infections, Wisconsin 1957; Minneapolis: 80–87.
Bawden, F. C., 1959. Physiology of virus diseases. – A Rev. Pl. Physiol. 10: 239–256.
Bawden, F. C., 1964. Plant viruses and virus diseases. Ronald Press Comp. New York 4 ed.
Bawden, F. C. & F. M. L. Sheffield, 1939. The intracellular inclusions of some plant virus diseases. – Ann. appl. Biol. 26: 102–115.
Beakbane, A. B. & F. C. Thompson, 1945. Abnormal lignification in the wood of some apple trees. – Nature (Lond.) 156: 145.
Beale, H. Purdy, 1937. Relation of Stanley's crystalline tobacco virus protein to intracellular crystalline deposits. – Contr. Boyce Thompson Inst. Pl. Res. 8: 413–431.
Beemster, A. B. R., 1958. Some aspects of mature plant resistance to viruses in the potato. – Proc. 3rd Conf. Potato Virus Diseases Lisse-Wageningen 1957: 212–217.
Benda, G. T. A., 1962. Heat-induced variegation – a model disease. – Phytopathology 52: 1307–1308.
Bennett, C. W., 1944. Latent virus of dodder and its effect on sugar beet and other plants. – Phytopathology 34: 77–91.
Bennett, C. W., 1952. Vein clearing in sugar beet induced by the beet leafhopper. – Phytopathology 42: 535–537.
Bennett, C. W., 1958. Masted plant viruses. – Symp. VII Int. Congr. Microbiol.: 218–223.
Bennett, C. W., 1960. Sugar beet yellows disease in the United States. – Tech. Bull. U.S. Dep. Agric. 1218.
Bercks, R., 1951. Weitere Untersuchungen zur Frage der Altersresistenz der Kartoffelpflanzen gegen das X-virus. – Phytopath. Z. 18: 249–269.
Beute, M. K. & J. L. Lockwood, 1968. Mechanism of increased root rot in virus-infected peas. – Phytopathology 58: 1643–1651.
Biraghi, A., 1952. Ulteriore contributo alla conoscenza del 'nanismo ruvido' del mais (With a summary: Further contributions to the knowledge of 'rough dwarf' of maize). – Ann. Sper. agraria 6: 1043–1053.
Black, L. M., 1945. A virus tumor disease of plants. – Am. J. Bot. 32: 408–415.
Black, L. M., 1946. Plant tumours induced by the combined action of wounds and virus. – Nature (Lond.) 158: 56–57.

Black, L. M., 1952. Plant virus tumors. – Ann. N.Y. Acad. Sci. 54: 1067–1075.

Black, L. M., 1965. Physiology of virus-induced tumors in plants. In W. Ruhland, ed.: Handbuch der Pflanzenphysiologie. – Springer Verl. Berlin, Heidelberg, New York XV/2: 236–266.

Black, L. M., C. Morgan & R. W. G. Wyckoff, 1950. Visualization of tobacco mosaic virus within infected cells. – Proc. Soc. exp. Biol. Med. 73: 119–122.

Black, L. M. & W. C. Price, 1940. The relationship between viruses of potato calico and alfalfa mosaic. – Phytopathology 30: 444–447.

Blakeslee, A. F., 1921. A graft-infectious disease of *Datura* resembling a vegetative mutation. – J. Genet. 11: 17–36.

Blattný, C., 1961. Virocecidien und Bemerkungen über die Taxonomie bei *Stellaria media* (L.) Vill. – Preslia 33: 206–208.

Blencowe, J. W. & T. Caldwell, 1949. Aspermy, a new virus disease of the tomato. – Ann. appl. Biol. 36: 320–322.

Bloch, R., 1954. Abnormal plant growth. – Brookhaven Symp. Biol. 6: 41–54.

Blodgett, E. C., 1944. Peach calico. – Phytopathology 34: 650–657.

Blodgett, E. C., J. A. Milbrath, E. L. Reeves & S. M. Zeller, 1951. Wart. – Agric. Handb. U.S. Dep. Agric. 10: 56–58.

Blumer, S., 1956. Über die Flachästigkeit (Rillenkrankheit) bei Apfelbäumen. – Schweiz. Z. Obst- u. Weinbau 65: 148–153.

Bode, O., 1968. Kartoffel. In M. Klinkowski, ed.: Pflanzliche Virologie II. Die Virosen des europäischen Raumes I. 2. Aufl. p. 31–68. Akademie Verlag, Berlin.

Boer, W. den, 1958. Beschadiging door natriumchloraat (with a summary: Damage due to sodium chlorate). – Tijdschr. Plziekt. 64: 144–149.

Böning, K., 1931. Zur Ätiologie der Streifen- und Kräuselkrankheit des Tabaks. – Z. ParasitKde 3: 103–141.

Bos, L., 1957. Heksenbezemverschijnselen, een pathologisch-morfologisch onderzoek (With a summary: Witches' broom phenomena, a pathomorphological study). – Meded. LandbHoogesch. Wageningen 57: 1–79.

Bos, L., 1957. Plant teratology and plant pathology. – Tijdschr. PlZiekt. 63: 222–231.

Bos, L., 1960. A witches' broom virus disease of *Vaccinium myrtillus* in the Netherlands. – Tijdschr. PlZiekt. 66: 259–263.

Bos, L., 1963. Heksenbezems en heksenbezemgroei by houtige gewassen (With a summary: Witches' brooms and witches' broom growth in woody plants.) 22e Jaarb. ned. dendrol. Vereen. 22 (1959, 1960, 1961): 162–174.

Bos, L., 1965a. Actieve afweer en ziekteverloop in met virus geïnfecteerde planten (With a summary: Active defense and course of the disease in virus-infected plants). – Meded. LandbHoogesch. OpzoekStns Gent 30: 1751–1758.

Bos, L., 1965b. Virussen en planten. – Tjeenk Willink, Zwolle.

Bos, L., 1966. De Vries i pervie chagi virusologii (De Vries and early virology). – O prirode virusov. Ind. 'Nauka' Moskwa, Proc. Conf. 100th Birthday Iwanowsky, Moscow Oct, 1964: 52–54.

Bos, L., 1967. Methods of studying plants as virus hosts. In: K. Maramorosch & H. Koprowsky, eds: Methods in virology. – Acad. Press Inc., New York, I: 129–162.

Bos, L., 1969. Inclusion bodies of bean yellow mosaic virus, some less known closely related viruses, and beet mosaic virus. – Neth. J. Pl. Path. 75: 137–143.

Bos, L., 1970. Plant morphogenesis and virus infection. – Miscell. papers 6 LandbHogesch. Wageningen: 55–64.

Bos, L., 1976. Problems and prospects in plant virus identification. – EPPO Bull. 6: 63–90.

Bos, L., 1977. Seed-borne viruses. In: W. B. Hewitt & L. Chiarappa eds. Plant health and quarantine in international transfer of genetic resources. – CRC Press, Inc., Cleveland, Ohio, 39–69.

Bos, L., B. Delević & J. P. H. van der Want, 1959. Investigations on white clover mosaic virus. – Tijdschr. PlZiekt. 65: 89–106.

Bos, L. & P. Grancini, 1965. Some experiments and considerations on the identification of witches' broom viruses, especially in clovers, in the Netherlands and Italy. – Neth. J. Pl. Path. 71:

Bos, L. & P. Grancini, 1968. Peculiar histoid enations in white clover and their relationships to virus diseases and toxic effects of leafhopper feeding. – Phytopath. Z. 61: 253–272.

Bos, L., K. Lindsten & D. Z. Maat, 1977. Similarity of clover yellow vein virus and pea necrosis virus. – Neth. J. Pl. Path. 83: 97–108.

Bos, L. & D. Z. Maat, 1965. A distinctive strain of the red clover mottle virus in the Netherlands. – Neth. J. Pl. Path. 71: 8–13.

Bos, L. & M. Rubio-Huertos, 1969. Light and electron microscopy of cytoplasmic and unusual nuclear inclusion bodies evoked by a virus from necrotic peas. – Virology, 37: 377–385.

Bos, L. & J. P. H. van der Want, 1962. Early browning of pea, a disease caused by a soil-and seed-borne virus. – Tijdschr. PlZiekt. 68: 368–390.

Bose, R. D. & S. D. Misra, 1938. Studies in Indian fibre plants, no. 6: Phyllody and some other abnormalities in the flower of sunn-hemp. – Indian J. agric. Sci. 8: 417–423.

Braak, J. P., 1955. Effect of some environmental factors on the appearance of 'June yellows' in strawberries and its significance for the development of a test method. – Euphytica 4: 189–166.

Brakke, M. K., A. E. Vatter & L. M. Black, 1954. Size and shape of wound-tumor virus. – Brookhaven Symp. Biol. 6: 137–156.

Brandes, J., 1957. Eine elektronenmikroskopische Schnellmethode zum Nachweis faden- und stäbchenförmiger Viren, insbesondere in Kartoffendunkelkeimen. – NachrBl. dt. PflSchutzdienst, Berl. 9: 151–152.

Brandes, J. & C. Wetter, 1963. Untersuchungen über Eigenschaften und Verwandtschaftsbeziehungen des Latenten *Passiflora*-Virus (*Passiflora* latent virus). – Phytopath. Z. 49: 61–70.

Brehmer, W. von & E. Rochlin, 1931. Histologische und mikrochemische Untersuchungen über pathologische Gewebeveränderungen viruskranker Kartoffelstauden. – Phytopath. Z. 3: 471–498.

Brenner, S. & R. W. Horne, 1959. A negative staining method for high resolution electron microscopy of viruses. – Biochim. biophys. Acta 34: 103–110.

Brierley, P. & F. P. McWhorter, 1936. A mosaic disease of iris. – J. agric. Res. 53: 621–635.

Brierley, P. & F. F. Smith, 1951. Survey of virus diseases of chrysanthemums. – Pl. Dis. Reptr. 35: 524–526.

Broadbent, L., 1964. The epidemiology of tomato mosaic. VII. The effect of TMV on tomato fruit yield and quality under glass. – Ann. appl. Biol. 54: 209–224.

Bruyn Ouboter, M. P. de, 1952. A new potato virus. – Proc. Conf. Potato Virus Diseases Wageningen-Lisse 1951: 83–84.

Butler, E. J. & S. G. Jones, 1949. Plant Pathology. - McMillan & Co Ltd. London.

Caldwell, J. & A. L. James, 1938. An investigation into the 'stripe' disease of narcissus. I. The nature and significance of the histological modifications following infection. – Ann. appl. Biol. 25: 244–253.

Calvert, E. L. & B. D. Harrison, 1966. Potato mop-top, a soil-borne virus. – Pl. Pathol. 15: 134–139.

Canova, A., 1963. Apple ring spot. – Tech. Commun. Commonw. Bur. Hort. Plantn Crops, 30: 37–40.

Carpenter, T. B., E. C. Calavan & D. W. Christiansen, 1965. Occurrence of excessive seed abortion in citrus fruits affected with stubborn disease. – Pl. Dis. Reptr. 49: 668–672.

Carter, W., 1973. Insects in relation to plant disease. – Interscience Publishers, New York, London, Sydney, Toronto, 2nd ed.

Casper, R., 1968. Blue light promotes symptom formation and virus multiplication in plants. – Virology 34: 179–181.

Čech, M. & H. Branišová, 1976. Nonrelatedness between symptoms and cucumber virus 4 content in different cucumber cultivars. – Biol. Plant. (Praha) 18: 58–62.

Cervantes, J. & R. H. Larson, 1961. Alfalfa mosaic virus in relation to tuber necrosis in the potato variety Red La Soda. – Res. Bull. agric. Exp. Stn Univ. Wis. 229: 1–40.

Cheo, P. C. & G. S. Pound, 1952. Relation of air temperature, soil temperature, photoperiod and light intensity to the concentration of cucumber virus 1 in spinach. – Phytopathology 42: 306–310.

Christie, R. G., 1967. Rapid staining procedures for differentiating plant virus inclusions in epidermal strips. – Virology 31: 268–271.

Christie, R. G. & J. R. Edwardson, 1977. Light and electron microscopy of plant virus inclusions. – Fla. Agric. Exp. Stns Monogr. Series 9: 150 pp.

Clinch, P., J. B. Loughnane & P. A. Murphy, 1936. A study of the aucuba or yellow mosaic of the potato. – Scient. Proc. R. Dubl. Soc. 21: 431–448.

Clusius, C., 1576. Rariorum aliquot stirpium per Hispanias observatarum historia Antverpiae: 512–515.

Cole, J. R., 1937. Bunch disease of pecans. – Phytopathology 27: 604–612.

Cole, R. M., J. G. Tully, T. J. Popkin & J. M. Bové, 1973. Morphology, ultrastructure, and bacteriophage infection of the helical mycoplasma-like organism (*Spiroplasma citri* gen. nov., sp. nov.) cultured from "stubborn" disease of citrus. – J. Bacteriol. 115: 367–386.

Corte, A. & A. C. Montemartini, 1965. Preliminary report on the influence of virus infection on the development of certain fungus and bacteria diseases in peach and apple trees. – Zašt. Bilja 85–88: 329–333.

Cos, P., 1637. Verzameling van een meenigte Tulipaanen. Naar het leven getekend met hunne Naamen en Swaarte der Bollen, zoo als die publieq verkogt zijn te Haarlem in den jaare A° 1637.

Costa, A. S., 1956. Anthocyanosis, a virus disease of cotton in Brazil. – Phytopath. Z. 28: 167–186.

Costa, A. S., 1965. Three whitefly-transmitted virus diseases of bean in Sao Paulo, Brazil. – FAO Pl. Prot. Bull. 13: 121–130.

Costa, A. S. & A. M. Carvalho, 1961. Studies on Brazilian tobacco streak. – Phytopath. Z. 42: 113–138.

Cronshaw, J., L. L. Hoefert & K. Esau, 1966. Ultrastructural features of *Beta* leaves infected with beet yellows virus. – J. Cell Biol. 31: 429–443.

Cropley, R., 1960. Pear blister canker, a virus disease. – Rep. E. Malling Res. Stn 1959: 104.

Davis, R. E., R. F. Whitcomb & R. L. Steere, 1968. Remission of aster yellows disease by antibiotics. – Science 161: 793–795.

Davis, R. E. & J. F. Worley, 1973. Spiroplasma; motile, helical microorganism associated with corn stunt disease. – Phytopathology 63: 403–408.

Desjardins, P. R., 1966. Inclusion bodies in *Nicotiana tabacum* produced by two strains of alfalfa mosaic virus. – Phytopathology 56: 875.

Desjardins, P. R., 1969. Alfalfa mosaic virus-induced lesions on bean: effect of light and temperature. – Pl. Dis. Reptr. 53: 30–33.

Diener, T. O., 1963. Physiology of virus-infected plants. – Rev. Phytopath. 1: 197–218.

Docters van Leeuwen, W. M., 1959. Nieuwe Nederlandse gallen: eerste supplement op het Gallenboek (With a summary: New galls from the Netherlands: first supplement to the 'Gallenboek'). – Ent. Ber., Amst. 19: 113–122.

Doi, Y., M. Teranaka, K. Yora & H. Asuyama, 1967. Mycoplasma- or PLT group-like microorganisms found in the phloem elements of plants infected with mulberry dwarf, potato witches' broom, aster yellows, or *Paulownia* witches' broom (Japanese with English summary). – Ann. phytopath. Soc. Japan 33: 259–266.

Dorst, H. J. van & D. Peters, 1974. Some biological observations on pale fruit, a viroid-incited disease of cucumber. – Neth. J. Pl. Path. 80: 85–96.

Duffus, J. E., 1977. Aphids, viruses and the yellow plague. In: K. Harris & K. Maramorosch, eds. Aphids as virus vectors. – Acad Press Inc. New York: 361–383.

Duggar, B. M. & J. K. Armstrong, 1923. Indications respecting the nature of the infective particles in the mosaic disease of tobacco. – Rep. Mo. bot. Gdn 10: 191–212.

Edwards, E. T., 1936. The witches' broom disease of lucerne. – Sci. Bull. Dep. Agric. N.S.W. 52.

Edwardson, J. R., 1974. Some properties of the potato virus Y-group. – Fla. Agric. Exp. Stns Monogr. Series 4: 398 pp.

Edwardson, J. R. & M. K. Corbett, 1961. A sexual transmission of cytoplasmic male sterility. – Proc. natn. Acad. Sci. U.S.A. 47: 390–396.

Edwardson, J. R. & M. K. Corbett, 1962. A virus-like syndrome in tomato caused by a mutation. – Am. J. Bot. 49: 571–575.

Edwardson, J. R. & D. E. Purcifull, 1970. Turnip mosaic virus-induced inclusions. – Phytopathology 60: 85–88.

Edwardson, J. R., D. E. Purcifull & R. G. Christie, 1968. Structure of cytoplasmic inclusions in plants infected with rod-shaped viruses. – Virology 34: 250–263.

Eibner, R., 1959. Untersuchungen über die 'Eisenfleckigkeit' der Kartoffel. – Diss. Giessen.

Elbertsen, M. & C. N. Silver, 1973. Nieuwe inzichten bij het herkennen van door tulpemozaïekvirus aangetaste tulpen. – Weekbl. Bloembollenteelt 83: 966–967.

Esau, K., 1933. Pathological changes in the anatomy of leaves of sugarbeet, *Beta vulgaris* L., affected by curly top. – Phytopathology 23: 679–712.

Esau, K., 1938. Some anatomical aspects of plant virus disease problems. – Bot. Rev. 4: 548–579.

Esau, K., 1944. Anatomical and cytological studies on beet mosaic. – J. agric. Res. 69: 95–117.

Esau, K., 1948a. Some anatomical aspects of plant virus disease problems II. – Bot. Rev. 14: 413–449.

Esau, K., 1948b. Anatomic effects of the viruses of Pierce's disease and phony peach. – Hilgardia 18: 423–482.
Esau, K., 1956. An anatomist's view of virus diseases. – Am. J. Bot. 43: 739–748.
Esau, K., 1957. Phloem degeneration in *Gramineae* affected by the barley yellow-dwarf virus. – Am. J. Bot. 44: 245–251.
Esau, K., 1960a. Cytologic and histologic symptoms of beet yellows. – Virology 10: 73–85.
Esau, K., 1960b. The development of inclusions in sugar beets infected with the beet-yellows virus. – Virology 11: 317–328.
Esau, K., 1961. Plants, viruses, and insects. – Harvard University Press, Cambridge Mass.
Esau, K., 1967. Anatomy of plant virus infections. – Rev. Phytopath. 5: 45–70.
Esau, K., 1968. Viruses in plant hosts. – University of Wisconsin Press, Madison, Milwaukee & London.
Esau, K. & J. Cronshaw, 1967. Relation of tobacco mosaic virus to the host cells. – J. Cell Biol. 33: 665–678.
Esau, K., J. Cronshaw & L. L. Hoefert, 1967. Relation of beet yellows virus to the phloem and to translocation in the sieve tube. J. Cell Biol. 32: 71–87.
Farkas, G. L. & F. Solymosy, 1965. Host metabolism and symptom production in virus-infected plants. – Phytopath. Z. 53: 85–93.
Farley, J. D. & J. L. Lockwood, 1964. Increased susceptibility to root rots in virus-infected peas. – Phytopathology 54: 1279–1280.
Fawcett, H. S. & A. A. Bitancourt, 1943. Comparative symptomatology of psorosis varieties in California. – Phytopathology 33: 837–864.
Feenstra, W. J., B. L. Johnson, P. Ribereau-Gayon & T. A. Geissman, 1963. The effect of virus infection on phenolic compounds in flowers of *Matthiola incana* R.Br. – Phytochemistry 2: 273–279.
Fluiter, H. J. de & N. Hubbeling, 1955. Waarnemingen over topvergeling bij erwten (With a summary: Observations on top yellows of peas). – Tijdschr. PlZiekt. 61: 165–175.
Folsom, D., W. C. Libby, G. W. Simpson & O. L. Wyman, 1938. Net necrosis of the potato. – Ext. Serv. Bull. Univ. Maine 246.
Ford, R. E., 1964. Pea pod streak symptoms correlated wtih time of virus infection. – Pl. Dis. Reptr. 48: 564–506.
Foster, W. R., T. B. Lott & M. F. Welsh, 1951. Little cherry. – In: U.S.D.A. Agriculture Handbook 10: 126–129.
Francki, R. I. B., 1967. Effect of high light intensities on spontaneous and virus-induced local lesions in *Gomphrena globosa*. – Phytopathology 57: 329.
Francki, R. I. B., 1973. Plant rhabdoviruses. – Adv. Virus Res. 18: 257–345.
Frazier, N. W., 1951. Red-leaf: an aphid-borne virus disease of filaree. Phytopathology 41: 221–227.

Freeman, J. A., H. A. Daubeny & R. Stace-Smith, 1969. Increased pollen abortion caused by viruses in four red raspberry cultivars. – Can. J. Pl. Sci. 49: 373–374.

Fridlund, P. R., 1970. Temperature effects on virus disease symptoms in some *Prunus*, *Malus* and *Pyrus* cultivars. – Wash. Agric. Exp. Sta. Bull. 726: 6 pp.

Fujisawa, I., M. Rubio-Huertos, C. Matsui & A. Yamaguchi, 1967. Intracellular appearance of cauliflower mosaic virus particles. – Phytopathology 57: 1130–1132.

Fukushi, T., E. Shikata & I. Kimura, 1962. Some morphological characters of rice dwarf virus. – Virology 18: 192–205.

Fukushi, T., E. Shikata, I. Kimura & M. Nemoto, 1960. Electron microscopic studies on the rice dwarf of virus. – Proc. Japan Acad. 36: 352–357.

Gäumann, E., 1946. Pflanzliche Infektionslehre. Verl. Birkhauser, Basel.

Gerola, F. M. & M. Bassi, 1966. An electron microscopy study of leaf vein tumours from maize plants experimentally infected with maize rough dwarf virus. – Caryologia, 19: 13–40.

Gibson, R. W., 1974. The induction of top-roll symptoms in potato plants by the aphid *Macrosiphum euphorbiae*. – Ann. appl. Biol. 76: 19–26.

Gifford, E. M., W. B. Hewitt, A. D. Graham & C. H. Lamoureux, 1956. An internal symptom for identifying fanleaf in the grapevine. – Bull. agric. Dep. Calif. 45: 268–272.

Goheen, A. C., G. Nyland & S. K. Lowe, 1973. Association of a Rickettsialike organism with Pierce's disease of grapevine and alfalfa dwarf and heat therapy of the disease in grapevines. – Phytopathology 63: 341–345.

Goheen, A. C. & W. C. Schnathorst, 1961. Resistance to powdery mildew in leafroll-affected grapevines. – Pl. Dis. Reptr. 45: 641–643.

Gold, A. H. & G. Faccioli, 1968. Relationship of ethylene to curly top symptoms. – Phytopathology 58: 1052.

Goldin, M. I., 1963. Virus inclusions in the plant cell and the nature of viruses (in Russian). – Acad. Sci. Moscow USSR.

Goldin, M. I., 1966. The physical action of viruses on the plant cell. – Proc. int. Conf. Pl. Viruses, Wageningen, July 1965: 158–165.

Goldstein, B., 1924. Cytological study of living cells of tobacco plants affected with mosaic disease. – Bull. Torrey bot. Club 51: 261–272.

Grancini, P., 1958. I sintomi del 'nanismo ruvido' del mais, (With a summary: Symptomatology of the maize dwarf disease (Rauhzwergkrankheit) in Italy – Maidyca 3: 67–79.

Griggs, W. H., D. D. Jensen & B. T. Iwakiri, 1968. Development of young pear trees with different rootstocks in relation to *Psylla* infestation, pear decline, and leaf curl. – Hilgardia 39: 153–204.

Grogan, R. G. & J. C. Walker, 1948. The relation of common mosaic to black root of bean. – J. agric. Res. 77: 315–331.

Hagedorn, D. J., L. Bos & J. P. H. van der Want, 1959. The red clover vein-mosaic virus in The Netherlands. – Tijdschr. PlZiekt. 65: 13–23.
Hagedorn, D. J. & J. C. Walker, 1949. Wisconsin pea stunt, a newly described disease. – J. agric. Res. 78: 617–626.
Harrison, B. D. & I. M. Roberts, 1968. Association of tobacco rattle virus with mitochondria. – J. gen. Virol. 3: 121–124.
Harrison, B. D., Z. Štefanac & I. M. Roberts, 1970. Role of mitochondria in the formation of X-bodies in cells of *Nicotiana clevelandii* by tobacco rattle virus. – J. gen. Virol. 6: 127–140.
Hartley, C. & F. W. Haasis, 1929. Brooming disease of black locust (*Robinia pseudacacia*). – Phytopathology 19: 163–166.
Hartmann, H. T. & D. E. Kester, 1968. Plant propagation, principles and practices. 2nd ed. – Prentice-Hall Inc., Englewood Cliffs, New Yersey.
Helms, K. & G. A. McIntyre, 1962. Studies on size of lesions of tobacco mosaic virus on Pinto bean. – Virology 18: 535–545.
Henke, O., 1957. Untersuchungen über einige am KH-Stoffwechsel und KH-Transport beteiligten Fermente in vergilbungskranken Zuckerrübenblättern. – Phytopath. Z. 29: 469–480.
Herrero, J., 1951. Studies of compatible and incompatible graft combinations with special reference to hardy fruit trees. – J. hort. Sci. 26: 186–237.
Hewitt, W. B. & E. M. Gifford, 1956. Symptoms for identifying fanleaf in dormant grapevines. – Bull Calif. Dept. Agric. 45: 249–252.
Hiebert, E. & J. G. McDonald, 1973. Characterization of some proteins associated with viruses in the potato Y group. – Virology 56: 349–361.
Hiebert, E., D. E. Purcifull, R. G. Christie & S. R. Christie, 1971. Partial purification of inclusions induced by tobacco etch virus and potato virus Y. – Virology 43: 638–646.
Hilborn, M. T., F. Hyland & R. C. McCrum, 1965. Pathological anatomy of apple trees affected by stem-pitting virus. – Phytopathology 55: 34–39.
Hildebrand, E. M., 1958. Masked virus infection in plants. – Rev. Microbiol. 12: 441–467.
Hill, A. V., 1943. Insect transmission and host plants of virescence (big bud of tomato). – J. Coun. sci. ind. Res. Aust. 16: 85–90.
Hitchborn, J. H. & G. J. Hills, 1965. The use of negative staining in the electron microscopic examination of plant viruses in crude extracts. – Virology 27: 528–540.
Hitchborn, J. H. & G. J. Hills, 1968. A study of tubes produced in plants infected with a strain of turnip yellow mosaic virus. – Virology 35: 50–70.
Hoefert, L. L. & E. M. Gifford, 1967a. Trabaeculae in the grapevine infected with leafroll virus. – Am. J. Bot. 54: 257–261.

Hoefert, L. L. & E. M. Gifford, 1967b. Grapevine leafroll virus – history and anatomic effects. – Hilgardia 38: 403–426.
Hofferbert, W. & G. zu Putlitz, 1955. Neue Erkenntnisse und Erfahrungen über die Blattrollkrankheit der Kartoffel. – NachrBl. dt. PflSchutzdienst Braunschweig 7.
Hofferek, H., 1967a. Untersuchungen über den Stoffwechsel viruskranker Pflanzen. III. Veränderungen des Stichstoff und Kohlenhydratstoffwechsels bei *Phaseolus vulgaris* L. nach Infektion mit dem Tabaksnekrose-Virus (TNV). (With a summary). – Phytopath. Z. 58: 141–174.
Hofferek, H., 1967b. Untersuchungen über den Stoffwechsel viruskranker Pflanzen. IV. Beeinflussung der Stoffwechselveränderungen bei *Phaseolus vulgaris* L. nach Infektion mit dem Tabaknekrose-Virus (TNV) durch Nährstoffapplikation (With a summary). – Phytopath. Z. 58: 214–238.
Hofferek, H. & H. Wolffgang, 1964. Untersuchungen über Stoffwechselveränderungen in virusinfizierten Pflanzen. (With summary). – Proc. Symp. host parasite relations in plant pathology, Budapest, 1964: 73–88. (TagBer. dt. Akad. LandwWiss. Berl. 74, 1965: 49–70).
Holmes, F. O., 1929. Local lesions in tobacco mosaic. – Bot. Gaz. 87: 39–55. Contr. Boyce Thompson Inst. 1: 504–520.
Holmes, F. O., 1931. Local lesions of mosaic in *Nicotiana tabacum* L. – Contr. Boyce Thompson Inst. 3: 163–172.
Holmes, F. O., 1964. Symptomatology of viral diseases in plants. – In M. K. Corbett & H. D. Sisler, eds: Plant virology, p. 17–38. Univ. Fla. Press Gainesville, Fla.
Hoof, H. A. van, 1952. Stip in kool, een virusziekte. (With a summary: 'Stip' (specks) in cabbage, a virus disease). – Meded. Dir. Tuinb. 15: 727–742.
Hoof, H. A. van, 1954. De oorzaak van 'koffiedik' in rode bewaarkool. – Zaadbelangen 8: 36.
Hooper, G. R. & H. Schneider, 1969. The anatomy of tumors induced on citrus by citrus vein-enation virus. – Am. J. Bot. 56: 238–247.
Hubbeling, N., 1954. Een virus als oorzaak van de zogenaamde 'voetziekte' bij erwten. – Zaadbelangen 14: 181–183.
Hubbeling, N., 1972. Resistance in beans to strains of bean common mosaic virus. – Meded. Fak. Landbouwwetensch. Gent 37: 458–466.
Hull, R., G. J. Hills & A. Plaskitt, 1969. Electron microscopy on *in vivo* aggregation forms of a strain of alfalfa mosaic virus. – J. Ultrastruct. Res. 26: 465–479.
Hutton, E. M. & N. E. Grylls, 1956. Legume 'little leaf' a virus disease of subtropical pasture species. – Aust. J. agric. Res. 7: 85–97.
Inouye, T., 1962. Studies on barley stripe mosaic in Japan. – Ber. Ōhara Inst. landw. Biol. 11: 413.
Isaacs, A., & J. Lindenmann, 1957. Virus interference I. The interferon. – J. Proc. R. Soc., B. 149: 258–267.

Ivancheva-Gabrovska, T., 1974. Recovering of tobacco plants infected by the tomato spotted wilt virus (Bulg. with Engl. Summ.) – In: I. Kovachevsky, D. Trifonov & M. Markov eds. Plant Virus Diseases. Publ. House Bulg. Acad. Sci.: 5–12.
Iwanowski, D., 1903. Über die Mosaikkrankheit der Tabakspflanze. Z. PflKrankh. – PflPath. PflSchutz 13: 1–41.
Jenkins, J. E. E. & I. F. Storey, 1955. Star cracking of apples in East Anglia – Pl. Path. 4: 50–52.
Jenkins, W. A., 1941. A histological study of snap bean tissues affected with black root. – J. Agric. Res. 62: 683–690.
Jensen, D. D., 1951. Mosaic or black streak disease of *Cymbidium* orchids. – Phytopathology 41: 401–414.
Jensen, D. D., W. H. Griggs, C. Q. Gonzales & H. Schneider, 1964. Pear decline virus transmission by pear psylla. – Phytopathology 54: 1346–1351.
Jensen, J. H., 1933. Leaf enations resulting from tobacco mosaic infection in certain species. of *Nicotiana* L. – Contr. Boyce Thompson Inst. Pl. Res. 5: 129–142.
John, V. T. & M. Weintraub, 1966. Symptoms resembling virus infection induced by high temperature in *Nicotiana glutinosa*. – Phytopathology 56: 502–506.
Johnson, E. M., 1930. Virus diseases of tobacco in Kentucky. – Kentucky Agric. Bull Ky agric. Exp. Stn 306: 289–415.
Johnson, J., 1925. Transmission of viruses from apparently healthy potatoes. – Res. Bull. agric. Exp. Stn. Univ. Wis. 63.
Juretić, N., 1968. Leaf and flower deformations of *Datura stramonium* L. infected with cucumber mosaic virus. – (Yugoslavian with English summary). – Acta bot. croat. 26, 27: 117–144.
Kahn, R. P. & R. L. Monroe, 1970. *Datura metel* L. as a virus-indicator plant. – Phytopathology 60: 1183–1185.
Kamei, T., M. Rubio-Huertos & C. Matsui, 1969. Thymidine-^{3}H uptake by X-bodies associated with cauliflower mosaic virus infection. – Virology 37: 506–508.
Kassanis, B., 1939. Intranuclear inclusions in virus-infected plants. – Ann. appl. Biol. 26: 705–709.
Kassanis, B., 1954. A virus latent in carnation and potato plants. – Nature (Lond.) 173: 1097.
Kassanis, B., 1956. Serological relationship between potato paracrinkle virus, potato virus S, and carnation latent virus. – J. gen. Microbiol. 15: 620–628.
Kassanis, B., 1957. Effects of changing temperature on plant virus diseases. – Adv. Virus Res. 4: 221–241.
Kassanis, B. & F. M. L. Sheffield, 1941. Variations in the cytoplasmic inclusions induced by three strains of tobacco mosaic virus. – Ann. appl. Biol. 23: 360–267.

Katwijk, W. van, 1955. Ruwschilligheid bij appels een virusziekte. – Tijdschr. PlZiekte. 61: 4–6.
Katwijk, W. van, 1956. Rough skin of apples. – Tijdschr. PlZiekt. 62: 46–49.
Kausche, G. A., E. Pfankuch & H. Ruska, 1939. Die Sichtbarmachung von pflanzlichem Virus im Übermikroskop. – Naturwissenschaften 27: 292–299.
Kegler, H., 1964a. Diagnose und Bekämpfung von Obstvirosen. – Sber. dt. Akad. LandwWiss. Berlin 13: 27–43.
Kegler, H., 1964b. Untersuchungen über Virosen des Kernobstes. VI. Die Flachästigkeit des Apfels. – Phytopath. Z. 50: 297–312.
Kehr, A. E. & H. H. Smith, 1954. Genetic tumors in *Nicotiana* hybrids. – Brookhaven Symp. Biol. 6(1953): 55–78.
Keitt, G. W. & C. N. Clayton, 1943. A destructive virus disease of sour cherry. – Phytopathology 33: 449–468.
Kellock, A. W., 1971. Red-leaf virus – a newly recognized virus disease of subterranean clover (*Trifolium subterraneum* L.). – Aust. J. Agric. Res. 22: 615–624.
Kerling, L. C. P., 1933. The anatomy of the 'kroepoek diseased' leaf of *Nicotiana tabacum* and of *Zinnia elegans*. – Phytopathology 23: 175–190.
Kienholz, J. R., 1939. Stony pit, a transmissible disease of pears. – Phytopathology 29: 260–267.
Kitajima, E. W., 1965. Electron microscopy of vira-cabeça virus (Brazilian tomato spotted wilt virus within the host cell). – Virology 26: 89–99.
Kitajima, E. W. & A. S. Costa, 1966. Morphology and development stages of *Gomphrena* virus. – Virology 29: 523–539.
Klinkenberg, C. H., 1940. Abnormale kurkvorming. Diss. Amsterdam.
Klinkowski, M., 1956. Beiträge zur Kenntnis der Virosen der Gladiole in Mitteldeutschland. – Mitt. BundAnst. Ld- u. Forstw. 85: 139–150.
Klinkowski, M., 1958. Beiträge zur Kenntnis der Stolburkrankheit der Kartoffel. – Proc. 3rd Conf. Potato Virus Diseases. Lisse-Wageningen 1957: 264–277.
Klinkowski, M. (ed.), 1967. Pflanzliche Virologie. I Einführung in die allgemeinen Probleme. – 2. Aufl. Akademie Verlag, Berlin.
Klotz, L. J. & E. C. Calavan, 1969. Gum diseases of citrus in California. – Calif. Agric. Exp. Stn. Ext. Circ. 396, 2nd rev.: 26 pp.
Knuhtsen, H., E. Hiebert & D. E. Purcifull, 1974. Partial purification and some properties of tobacco etch virus induced intranuclear inclusions. – Virology 61: 200–209.
Köhler, E., 1952. Die Bukettkrankheit, eine Viruskrankheit der Kartoffel. – Phytopath. Z. 19: 284–294.
Köhler, E., 1964. Pathologische Physiologie. In: E. Köhler, Allgemeine Viruspathologie der Pflanzen, p. 22–31. – Parey, Berlin und Hamburg.

Koot, Y. van & H. J. M. van Dorst, 1958. De invloed van enkele virussen op de bloemvorming bij komkommer (with English summary). – Tijdschr. PlZiekt. 64: 432–439.
Koot, Y. van & H. J. M. van Dorst, 1959. Virusziekten van komkommer in Nederland (with a summary: Virus diseases of cucumber in the Netherlands). – Tijdschr. PlZiekt. 65: 257–271.
Koshimizu, S. & T. Iizuka, 1957. Relationship between the brown speck of soybean seed and the soybean mosaic. – Ann. phytopath. Soc. Japan 22: 18.
Kovačevski, I. C., 1959. Eine stark nekrotische Variante des Bronzeflecken-krankheit-Virus an Tomaten in Nordbulgarien. – 5th Int. Congr. Crop Prot., Hamburg 1957: 371–394.
Kreitlow, K. W., Hunt, O. J. & H. L. Wilkins, 1957. The effect of virus infection on yield and chemical composition of Ladino clover. – Phytopathology 47: 390–394.
Kristensen, H. Rønde & A. Thomsen, 1958. Chrysanthemum viroser. Tidsskr. Pl. Avl. 62: 627–669.
Kunkel, L. O., 1924a. Insect transmission of aster yellows. – Pnytopathology 14: 54.
Kunkel, L. O., 1924b. Histological and cytological studies on the Fiji disease of sugar cane. – Bull Hawaiian Sug. Plrs' Ass. Exp. Stn, Bot. Ser. 3: 99–107.
Kunkel, L. O., 1931. Studies on aster yellows in some new host plants. – Contr. Boyce Thompson Inst. Pl. Res. 3: 85–123.
Kunkel, L. O., 1944. General pathology of virus infections in plants. – In: R. Doerr & C. Hallauer, eds: Handbuch der Virusforschung. I Ergänzungband, Wien: 473–521.
Kunkel, L. O., 1954a. Virus-induced abnormalities. – Brookhaven Symp. Biol. 6: 157–173.
Kunkel, L. O., 1954b. Maintenance of yellows-type viruses in plant and insect reservoirs. – The dynamics of virus and Rickettsial infections. – Int. Symp. New York: 150–163.
Küster, E., 1911. Die Gallen der Pflanzen. – Leipzig.
Küster, E., 1925. Pathologische Pflanzenanatomie. – 3e Aufl. Fischer Jena.
Larson, R. H., 1959. Purple top hair sprout and low soil temperature in relation to secondary or sprout tuber formation. – Am. Potato J. 36: 29–31.
Lawson, R. H., 1968. Cineraria varieties as starch lesion test plants for chrysanthemum stunt virus. – Phytopathology 58: 690–695.
Lawson, R. H. & S. Hearon, 1970. Subcellular localization of chrysanthemum aspermy virus in tobacco and chrysanthemum leaf tissue. – Virology 41: 30–37.
Lee, C. L., 1955. Virus tumor development in relation to lateral-root and bacterial-nodule formation in *Melilotus alba*. – Virology 1: 152–164.

Lee, C. L. & L. M. Black, 1955. Anatomical studies of *Trifolium incarnatum* infected by wound-tumor virus. – Am. J. Bot. 42: 160–168.

Leeuw, G. T. N. de, 1968. Translocation pathways of tobacco mosaic virus in *Nicotiana tabacum* L. var. Xanthi-nc. – Diss. Utrecht.

Lihnell, D., 1958. Investigations on spraing. – Proc. 3rd Conf. Potato Virus Diseases, Lisse-Wageningen 1957: 184–188.

Limberk, A. & M. Ulrychová, 1972. Vivipary in fruits of tomato plants infected with a mycoplasma disease – potato witches' broom. – Phytopath. Z. 73: 227–234.

Lindner, R. C., 1961. Chemical tests in the diagnosis of plant virus diseases. – Bot. Rev. 27: 501–521.

Lloyd, A. J. E., H. G. Smith & L. H. Jones, 1965. Evesham stain, a virus disease of broad beans (*Vicia faba* L.). – Hort. Res. 5: 13–18.

Loon, L. C. van, 1972. Pathogenese en symptoomexpressie in viruszieke tabak; een onderzoek naar veranderingen in oplosbare eiwitten. (with Engl. summ.: Pathogenesis and symptom expression in virus-diseased tobacco plants, a study of changes in soluble proteins) – Diss. Agric. Univ. Wageningen: 152 pp.

Loon, L. C. van, 1977. Induction by 2-chloroethylphosphonic acid of viral-like lesions, associated proteins, and systemic resistance in tobacco. – Virology 80: 417–420.

MacLeod, R., L. M. Black & F. A. Moyer, 1966. The fine structure and intracellular localization of potato yellow dwarf virus. – Virology 29: 540–552.

Maramorosch, K., 1953. A new leafhopper-borne plant disease from Western Europe. – Pl. Dis. Reptr. 37: 612–613.

Maramorosch, K., 1959. An ephemeral disease of maize transmitted by *Dalbulus elimatus*. – Entomologia exp. appl. 2: 169–170.

Marenaud, C., 1966. Incompatibilité de greffage induite par les viroses chez les arbres fruitiers. – Rev. Pathol. gén. comp. 66: 259–262.

Marrou, J. & A. Migliori, 1968. Variation périodique du pouvoir infectieux du virus de la mosaïque du concombre dans les feuilles de tabac infectées par le virus. – C. R. Acad. Sci. Paris 267: 625–628.

Marrou, J., A. Migliori & G. Marchoux, 1968. Variation dans le temps du pouvoir infectieux du virus de la mosaïque du concombre chez différentes espèces végétales. – Annls Épiphyt. 19: 83–92.

Martelli, G. P. & M. Russo, 1969. Nuclear changes in mesophyll cells of *Gomphrena globosa* L. associated with infection by beet mosaic virus. – Virology 38: 297–308.

Martelli, G. P. & M. Russo, 1977. Plant virus inclusion bodies. – Adv. Virus Res. 21: 175–266.

Martin, C., 1958. Anomalies de synthèse des anthocyanes dans le germe de pomme de terre atteinte de maladies à virus. – C. r. hebd. Séanc. Acad. Sci. Paris 246(1956): 2790–2792.

Martin, C. & J. Quemener, 1956. Sur un test colorimétrique et quelques symptomes permettant la détection des maladies à virus chez la pomme de terre. – C. r. hebd. Séanc. Acad. Agr. Fr. 42: 426–428.

Masters, M. T., 1869. Vegetable Teratology. – London 1869.

Masters, M. T., 1877. Action of scion on stock. – Gdnrs' Chron. 7: 730.

Matsui, C., & A. Yamaguchi, 1966. Some aspects of viruses in situ. – Adv. Virus Res. 12: 127–174.

Matthews, R. E. F., 1970. Effects on plant metabolism. In: R. E. F. Matthews: Plant Virology. – Acad. Press, New York and London 1970: 319–348.

Mayer, A., 1886. Über die Mosaikkrankheit des Tabaks. – Landw. Vers-Stnen 32: 450–467.

McKay, M. P. & M. F. Warner, 1933. Historical sketch of tulip mosaic or breaking. The oldest known plant virus disease. – Natn. hort. Mag., July: 187.

McWhorter, F. P., 1938. The antithetic virus theory of tulip-breaking. – Ann. appl. Biol. 25: 254–270.

McWhorter, F. P., 1941. Isometric crystals produced by *Pisum* virus 2 and *Phaseolus* virus 2. – Phytopathology 31: 760–761.

McWhorter, F. P., 1950. Neoplasms induced in plants by viruses. – In: Viruses. Oregon State College Biol. Colloquium: 14–15.

McWhorter, F. P., 1951. The examination of tissues in living leaves and flowers by means of a high vacuum technic. – Stain Technol. 26: 177–180.

McWhorter, F. P., 1965. Plant virus inclusions. – Rev. Phytopath. 3: 287–312.

Milbrath, J. A. & S. M. Zeller, 1945. Latent viruses in stone fruits. – Science (Lond.) 101: 114–115.

Miličić, D. & Z. Štefanac, 1971. Cell inclusions of the cucumber green mottle mosaic virus and the *Odontoglossum* ringspot virus. – Acta Bot. Croat. 30: 33–40.

Miličić, D., Z. Štefanac, N. Juretić & M. Wrischer, 1968. Cell inclusions of Holmes' Ribgrass virus. – Virology 35: 356–362.

Milne, R. G., 1972. Electron microscopy of viruses. – In: C. I. Kado & H. O. Agrawal, Principles and techniques in plant virology. Van Nostrand Reinhold Comp., New York, Cincinnati, Toronto, London, Melbourne: 76–128.

Mischke, W., 1966. Der 'Shirofugen'-Test zum Nachweis von Ringflecken-viren des Steinobstes. – NachrBl. dt. PflSchutzdienst., Stuttg. 18: 20–25.

Moericke, V., 1955. Über den Nachweis der Blattrollkrankheit in Kartoffelknollen durch den Resorzintest. – Phytopath. Z. 24: 462–464.

Molisch, H., 1885. Über merkwürdig geformte Proteinkörper in den Zweigen von *Epiphyllum*. – Ber. dt. bot. Ges. 3: 195.

Moore, J. D., 1967. Relation of temperature to prune dwarf symptom expression on Italian prune leaves. – Phytopathology 57: 823.

Morwood, R. B., 1954. Peanut diseases. – Qd agric. J. 79: 267–270.

Mosse, B., 1962. Graft-incompatibility in fruit trees. Techn. Comm. Commonw. Bur. Hort. Plant. Crops, East Malling 28: 36 pp.

Mulder, D., 1951. De Eckelrader virusziekte van zoete kersen. – Meded. Dir. Tuinb. 14: 217–228.

Mulder, D., 1953. De proliferatieziekte van appel, een virusziekte (With a summary: Proliferation disease of apple, a virus disease). – Tijdschr. PlZiekt. 59: 72–76.

Murant, A. F., J. Chambers & A. T. Jones, 1974. Spread of raspberry bushy dwarf virus by pollination, its association with crumbly fruit, and problems of control. – Ann. appl. Biol. 77: 271–281.

Murant, A. F. & I. M. Roberts, 1971. Cylindrical inclusions in coriander leaf cells infected with parsnip mosaic virus. – J. gen. Virol. 10: 65–70.

Murphy, P. A., 1923. On the cause of rolling in potato foliage; and on some further insect carriers of the leaf-roll disease. – Scient. Proc. R. Dubl. Soc. 17: 163–184.

Mwanza, M. P. & L. E. Williams, 1966. Viruses as predisposing factors in the susceptibility of corn and wheat plants to other pathogens. – Phytopathology 56: 892.

Newmark, P. & D. Fraser, 1956. Composition of an abnormal protein present in tobacco plants infected with tobacco mosaic virus. – J. Am. chem. Soc. 78: 1588–1590.

Nieuwhof, M., F. Garritsen & A. Kraai, 1974. Grey speck disease, a non-parasitic post-harvest disorder of storage white cabbage (*Brassica oleracea* var. *capitata* L.F. alba D.C.). – Euphytica 23: 1–10.

Nieuwhof, M. & D. Wiering, 1962. Heritable necrosis of red storage cabbage – Euphytica 11: 240–244.

Nishimura, M., 1918. A carrier of the mosaic disease. – Bull. Torrey bot. Club. 45: 219–233.

Noordam, D., 1952. Virusziekten bij chrysanten in Nederland (With a summary: Virus diseases of *Chrysanthemum indicum* in The Netherlands). – Tijdschr. PlZiekt. 58: 121–190. Meded. IPO 46.

Novák, J., 1964. Wachstumsmodifikationen und Bewegungsänderungen bei Tabakmosaikvirus-infizierten Tabakpflanzen. – Phytopath. Z. 51: 41–61.

Novák, J., 1975. Wachstumsänderungen der Tabakpflanze *Nicotiana tabacum* cv. Samsun nach der Inokulation des Tabakmosaikvirus. – Acta Sc. Nat. Brno 9 (8): 41 pp.

Nuorteva, P., 1962. Studies on the causes of the phytopathogenicity of *Calligypona pellucida* (F.) (Hom. Araeopidae). – Ann. Zool. Soc. 'Vanamo' 23: 1–58.

Nyland, G. & A. C. Goheen, 1969. Heat therapy of virus diseases of perennial plants. – Ann. Rev. Phytopath. 7: 331–354.
Oort, A. J. P., 1959. Over de termen primair en secondair ziek in de fytopathologie (With a summary: On the use of the terms 'primair' and 'secondair ziek' in plant pathology). – Tijdschr. PlZiekt. 65: 142–146.
Oosten, H. J. van, 1972. Diagnosis of sharka (plum pox) by internal and external fruit symptoms. – Neth. J. Pl. Path. 78: 99–106.
Oosten, H. J. van & C. H. J. van Bakel, 1970. Inclusion bodies in plants infected with sharka (plum pox) virus. – Neth. J. Pl. Path. 76: 313–319.
Opel, H., 1965. Untersuchungen zur Physiologie der Symptombildung viruskranker Pflanzen. – TagBer. dt. Akad. LandwWiss. Berl. 74: 83–97.
Opel, H., 1967. Der Einfluss der Bewurzelung auf Virusvermehrung und Symptombildung abgeschnittener Blätter. – Flora, Abt. A. 158: 157–163.
Opel, H., 1970. Einfluss des Lichtes auf Virusvermehrung und Symptombildung virusinfizierter Pflanzen. – Arch. Phytopath. Pflanzensch. 6: 303–319.
Opel, H., G. Wolf & H. Kegler, 1974. *Pseudomonas syringae* van Hall als Ursache virose-ähnlicher Symptome auf *Chenopodium quinoa* Willd. bei Versuchen zur Übertragung von Obstviren. – Arch. Phytopath. Pflanzensch. 10: 383–393.
Oshima, N. & M. F. Kernkamp, 1957. Effects of viruses on overwintering of red clover in Minnesota. – Pl. Dis. Reptr. 41: 10.
Oswald, J. W. & B. R. Houston, 1953. The yellow-dwarf virus disease of cerial crops. – Phytopathology 43: 128–136.
Pal, B. P. & P. Nath, 1935. Phyllody, a possible virus disease of *Sesamum*. – Indian J. Agric. Sci. 5: 517–522.
Paul, H. L., 1974. SDS polyacrylamide gel electrophoresis of virion protein as a tool for detecting the presence of virus in plants. – Phytopath. Z. 80: 330–339.
Paul, H. L., 1975. SDS polyacrylamide gel electrophoresis of virion proteins as a tool for detecting the presence of virus in plants. II Examination of further virus-host combinations. – Phytopath. Z. 83: 303–310.
Paul, H. L. & L. Quantz, 1959. Über den Wechsel der Konzentration des echten Ackerbohnenmosaikvirus in Ackerbohnen. – Arch. Mikrobiol. 32: 312–318.
Pennazio, S. & A. Appiano, 1975. Potato virus X aggregates in ultrathin sections of *Datura* meristematic dome – Phytopath. Mediterr. 14: 12–15.
Penzig, O., 1921–1922. Pflanzen-Terotologie. – 2e Aufl. Berlin.
Pine, T. S., 1963. Shock symptoms of the stone fruit ringspot virus complex in apricot. – Phytopathology 53: 179–183.
Plese, N., D. Miličić & D. Trifonov, 1971. Anatomy of pear trees affected by blister canker and bud drop. – Annls Phytopath. 3, no hors-série: 265–283.

Ploaie, P. & K. Maramorosch, 1969. Electron microscopic demonstration of particles resembling Mycoplasma or Psittacosis–Lymphogranuloma-Trachoma Group in plants infected with European yellows-type diseases. – Phytopathology 59: 536–544.

Porter, C. A., 1959. Biochemistry of plant virus infection. – Adv. Virus Res. 6: 75–91.

Posnette, A. F., 1947. Virus diseases of cacao in West Africa. I. Cacao viruses 1A, 1B, 1C and 1D. – Ann. appl. Biol 34: 388–402.

Posnette, A. F., 1953. Green petal – a new virus disease of strawberries. – Pl. Path. 2: 17–18.

Posnette, A. F. & R. Cropley, 1961. The effect of latent virus infection on the growth and yield of sweet cherry trees. – Tidsskr. PlAvl. 65: 118–124.

Posnette, A. F. & R. Cropley, 1963. Apple stem pitting. In: A. F. Posnette ed. Virus diseases of apples and pears. – Commonw. agric. Bur., Farnham Roy. Bucks. England 77–78.

Pound, G. S. & J. C. Walker, 1953. Autogenous necrosis of cabbage. – Phytopathology 43: 415–418.

Praceus, C., 1958. Anatomische Untersuchungen an Enationen von Aspermie-Virus infizierten Pflanzen. – Phytopath. Z. 33: 248–262.

Prentice, I. W., 1950a. Rubus-stunt, a virus disease. – J. hort. Sci. 26: 35–42.

Prentice, I. W., 1950b. Experiments on rubbery wood disease of apple trees. A progress report. – Rep. E. Malling Res. Stn 1949: 122–125.

Prentice, I. W., 1952. Resolution of strawberry virus complexes. V. Experiments with viruses 4 and 5. – Ann. appl. Biol. 39: 487–494.

Price, W. C., 1932. Acquired immunity to ring-spot in *Nicotiana*. – Contr. Boyce Thompson Inst. Pl. Res. 4: 359–403.

Price, W. C., 1936. Virus concentration in relation to acquired immunity from tobacco ring spot. – Phytopathology 26: 503–529.

Proeseler, G., 1968. Gallmilben verursachen virusähnliche Schadbilder. – NachrBl. dt. PflSchutzdienst, Berlin, 3: 48–52.

Purcifull, D. E. & D. L. Batchelor, 1977. Immunodiffusion tests with sodium dodecyl sulfate (SDS)-treated plant viruses and plant viral inclusions. – Univ. Fla. Agric. Exp. Sta. Techn. Bull. 788: 39 pp.

Purcifull, D. E., J. R. Edwardson & R. G. Christie, 1966. Electron microscopy of intracellular aggregates in pea (*Pisum sativum*) infected with clover yellow mosaic virus. – Virology 29: 276–284.

Quanjer, H. M., 1913. Die Nekrose des Phloëms der Kartoffelpflanze, die Ursache der Blattrollkrankheit. – Meded. Rijks Hoogere Land-, Tuin- en Boschbouwschool, 6: 41–80.

Quanjer, H. M., 1931. The methods of classification of plant viruses and an attempt to classify and name potato viroses. – Phytopathology 21: 577–613.

Quantz, L., 1958a. Ein Beitrag zur Kenntnis der Erbsenvirosen in Deutschland. – Nachr. Bl. dt. Pfl. SchutzDienst, Braunschweig 10: 67–70.
Quantz, L., 1958b. Untersuchungen zur Bestimmung mosaikresistenter, überempfindlicher Gartenbohnensorten (*Phaseolus vulgaris* L.) im Labortest. – Phytopath. Z. 31: 319–330.
Rademacher, B. & R. Schwarz, 1958. Die Rotblättrigkeit oder Blattröte des Hafers, eine Viruskrankheit (*Hordeumvirus nanescens*). – Z. PflKrankh. PflPath. Sch. 65: 641–650.
Ragetli, H. W. J., 1967. Virus-host interactions, with emphasis on certain cytopathic phenomena. – Can. J. Bot. 45: 1221–1234.
Rasa, E. A. & K. Esau, 1961. Anatomic effects of curly top and aster yellows viruses on tomato. – Hilgardia 30: 469–529.
Raymer, W. B. & J. A. Milbrath, 1960. The identity and host relations of the potato late-breaking virus. – Phytopathology 50: 312–319.
Reid, M. S. & R. E. F. Matthews, 1966. On the origin of the mosaic induced by turnip yellow mosaic virus. – Virology 28: 563–570.
Rich, S., 1949. Some relations between *Phaseolus* virus 2 and its associated crystalline inclusions. – Phytopathology 39: 221–224.
Richards, B. L. & E. L. Reeves, 1951. Necrotic rusty mottle. – U.S. D. A. Agriculture Handbook 10: 120–122.
Rivers, T. M., 1937. Viruses and Koch's postulates. – J. Bact. 33: 1–12.
Robb, S. M., 1963. A method for the detection of dahlia mosaic virus in *Dahlia*. – Ann. appl. Biol. 52: 145–148.
Robb, S. M., 1964. Location, structure and cytochemical staining reactions of the inclusion bodies found in *Dahlia variabiles* infected with *Dahlia* mosaic virus. – Virology 23: 141–144.
Roberts, D. A., 1956. Influence of alfalfa mosaic virus infection upon winter hardiness of Ladino clover. – Phytopathology 46: 24.
Roberts, I. M. & B. D. Harrison, 1970. Inclusion bodies and tubular structures in *Chenopodium amaranticolor* plants infected with strawberry latent ringspot virus. – J. gen. Virol. 7: 47–54.
Robertson, H. F., 1928. Rep. Mycol. Burma. (Abstr. in Rev. appl. Mycol. 8: 355).
Rosenzopf, E., 1951. Sind Eiweiszspindeln Virus-Einschlusskörper? – Phyton 3: 95.
Ross, A. F., 1941. The concentration of alfalfa mosaic virus in tobacco plants at different periods of time after inoculation. – Phytopathology 31: 410–420.
Ross, A. F., 1961a. Localized acquired resistance to plant virus infection in hypersensitive hosts. – Virology 14: 329–339.
Ross, A. F., 1961b. Systemic acquired resistance induced by localized virus infections in plants. – Virology 14: 340–358.
Ross, J. P., 1968. Effect of single and double infections of soybean mosaic and bean pod mottle viruses on soybean yield and seed characters. – Pl. Dis. Reptr 52: 344–348.

Roy, S. C., 1931. A preliminary note on the occurrence of sepaloidy and sterility in til (*Sesamum indicum*). – Agriculture Live-Stk India 1: 282–285.

Rozendaal, A. & J. H. Brust, 1955. The significance of potato virus S in seed potato culture. – Proc. Conf. Potato Virus Diseases, Lisse-Wageningen 1954: 120–133.

Rozendaal, A. & J. P. H. van der Want, 1948. Over de identiteit van het ratelvirus van de tabak en het stengelbontvirus van de aardappel (With a summary: On the identity of the rattle virus of tobacco and stem-mottle virus of potato). – Tijdschr. PlZiekt. 54: 113–133.

Rubio-Huertos, M., 1950. Estudios sobre inclusiones intracellulares producidas por virus en las plantas. – Microbiología esp. 3: 207–232.

Rubio-Huertos, M., 1954. Rapid extraction of intact crystalline inclusions from the cells of plants infected with tobacco mosaic virus. – Nature (Lond.) 174: 313.

Rubio (-Huertos), M., 1956. Origin and composition of cell inclusions associated with certain tobacco and crucifer viruses. – Phytopathology 46: 553–556.

Rubio-Huertos, M., 1968. Further studies on ultrastructure of plants infected with *Petunia* ringspot virus. – Protoplasma 65: 465–476.

Rubio-Huertos, M., 1972. Inclusion bodies. – In: C. I. Kado & H. O. Agrawal, eds. Principles and techniques in plant virology. Van Nostrand Reinhold Comp., New York, Cincinnati, Toronto, London, Melbourne: 62–75.

Rubio-Huertos, M. & L. Bos, 1969. Morphology and intracellular localization of bacilliform virus particles associated with the clover enation disease. – Neth. J. Pl. Path. 75: 329–337.

Rubio-Huertos, M. & L. Bos, 1973. Light and electron microscopy of red clover vein mosaic virus in pea *(Pisum sativum)*. – Neth. J. Pl. Path. 79: 94–103.

Rubio-Huertos, M., S. Castro, I. Fujisawa & C. Matsui, 1972. Electron microscopy of the formation of carnation etched ring virus intracellular inclusion bodies. – J. gen. Virol. 15: 257–260.

Rubio-Huertos, M., & F. G. Hidalgo, 1964. Ultrathin sections of intranuclear and intracytoplasmic inclusions induced by severe etch virus. – Virology 24: 84–90.

Russell, G. E., 1968. Specific interactions between virus and fungus diseases in sugar beet. – Int. Congr. Pl. Path. London: 168.

Russo, M. & G. P. Martelli, 1972. Ultrastructural observations on tomato bushy stunt virus in plant-cells. – Virology 49: 122–129.

Russo, M., G. P. Martelli & A. Quacquarelli, 1968. Studies on the agent of Artichoke mottle crinckle. IV. Intracellular localization of the virus. – Virology 34: 679–693.

Ryder, E. J., 1964. Transmission of common lettuce mosaic virus through the gametes of the lettuce plant. – Pl. Dis. Reptr 48: 522.

Sampson, P. J. & R. H. Taylor, 1968. A comparison of the electron microscope, microprecipitin tests and indicator plants for the detection of potato viruses S, X and Y. – Phytopathology 58: 489–493.
Samuel, G., 1931. Some experiments on inoculating methods with plant viruses and on local lesions. – Ann. appl. Biol. 18: 494–507.
Samuel, G., J. G. Bald & H. A. Pittman, 1930. Investigations on spotted wilt of tomatoes. – Bull. Coun. scient. ind. Res., Melb. 44:
Savulescu, A. & I. Pop, 1956. Contributii la studiul stolburuluc în Romînia (Avec résumé français: Contribution à l'étude du 'Stolbur' en Roumanie). – Buletin şti. Acad. Repub. pop. rom., Secţia de Biol. şi ştünţe agricole 8: 723–737.
Schippers, B., 1963. Transmission of bean common mosaic virus by seed of *Phaseolus vulgaris* L. cultivar Beka. – Acta bot. neerl. 12: 433–497.
Schlegel, D. E., S. H. Smith & G. A. de Zoeten, 1967. Sites of virus synthesis within cells. – Rev. Phytopath. 5: 223–246.
Schmelzer, K., 1967. Symptome des Gurkenmosaik-Virus an Kürbissen. – NachrBl. dt. PflSchutzdienst, Berl. 21: 109–110.
Schmelzer, K. & A. Molnár, 1964. *Pseudomonas aptata* (Brown et Jamieson) Stevens in Zusammenhang mit vermeintlichen pflanzlichen Virusinfektionen. – Phytopath. Z. 50: 112–118.
Schmelzer, K. & H. O. Schmidt, 1963a. Virusbedingt buntblättrige Ziergehölze – eine Gefahr für den Nutzpflanzenanbau. – Gartenwelt 63: 385–388.
Schmelzer, K. & H. O. Schmidt, 1963b. Merkmale virus- und genetisch bedingter Buntblättrigkeiten. – Gartenwelt 63: 405–407.
Schneider, H., 1954. Anatomy of bark of bud union, trunk, and roots of quick-decline-affected sweet orange trees on sour orange rootstocks. – Hilgardia 22: 567–581.
Schneider, H., 1959. The anatomy of tristeza-virus infected citrus. – Proc. Conf. Citrus Virus Dis. Riverside, Calif. 1957: 73–84.
Schneider, H., 1973. Cytological and histological aberrations in woody plants following infection with viruses, mycoplasmas, Reckettsias, and flagellates. – Annu. Rev. Phytopath. 11: 119–146.
Schnepf, E. & J. Brandes, 1961. Über ein Virus aus *Passiflora* spec. – Phytopath. Z. 43: 102–105.
Scholten, G., 1966. Een erfelijke afwijking bij de cyclamen-cultivar 'Sylphide' (With a summary: A genetical variation of the cyclamen cultivar 'Sylphide'). – Neth. J. Pl. Path. 72: 28–30.
Scholz, M., 1962. Die Bedeutung des S-Virus für den Kartoffelbau und Probleme der S-Virussanierung. (With a summary: The significance of virus S for seed-potato growing and problems of virus S sanitation). – NachrBl. dt. PflSchutzdienst, Berl. 16: 174–179.
Schossig, S., A. Kuhn & H. Kegler, 1965. Einfluss von Viruserkrankungen. auf den Verwachsungserfolg bei Okulationen. – Dt. Gärtner-Post 27.

Schuster, G., 1956. Zum Kallosetest ('Igel-Lange-Test') für den Virusnachweis an Kartoffeln. – NachrBl. dt. PflSchutzdienst, Berl. 10: 243–250.

Scurfield, D. G. & D. E. Bland, 1963. The anatomy and chemistry of 'rubbery' wood in apple var. Lord Lambourne. – J. hort. Sci. 38: 297–306.

Scurfield, G. & C. Reinganum, 1964. Some anatomical and chemical effects of 'flat limb' virus on apple var. Gravenstein. – Aust. J. agric. Res. 15: 548–559.

Sela, I., I. Harpaz & Y. Birk, 1966. Identification of the active component of an antiviral factor isolated from virus-infected plants. – Virology 28: 71–78.

Selman, I. W. & R. G. Milne, 1961. A new local lesion assay method for tomato spotted wilt virus, with a note on cyclical changes in infectivity. – Pl. Path. 10: 100–104.

Sequeira, O. A. de & R. Cropley, 1968. Symptoms in 'Virginia Crab' caused by the E 36 virus. – TagBer. VII. Europ. Symp. Viruskrankh. Obstbäume 1967: 35–42.

Severin, H. H. P., 1940. Potato, naturally infected with California aster yellows. – Phytopathology 30: 1049–1051.

Shalla, T. A., 1968a. Effects of viruses on cellular components. – Abstr. Papers First Intern. Congr. Pl. Pathol. 178.

Shalla, T. A., 1968b. Virus particles in chloroplasts of plants infected with the U5 strain of tobacco mosaic virus. – Virology 35: 194–203.

Shalla, T. A., L. Chiarappa & T. W. Carroll, 1963. A graft transmissible factor associated with pear decline. – Phytopathology 53: 366–367.

Shalla, T. A. & J. F. Shepard, 1972. The structure and antigenic analysis of amorphous inclusion bodies induced by potato virus X. – Virology 49: 654–667.

Shear, C. L., 1916. False blossom of the cultivated cranberry. – Bull. U.S. Dep. Agric. 444.

Sheffield, F. M. L., 1931. The formation of intracellular inclusions in solanaceous hosts infected with aucuba mosaic of tomato. – Ann. appl. Biol. 18: 471–493.

Sheffield, F. M. L., 1941. The cytoplasmic and nuclear inclusions associated with severe etch virus. – Jl R. microsc. Soc. 61: 30–45.

Sheffield, F. M. L., 1946. Preliminary studies in the electron microscope of some plant virus inclusion bodies. – Jl R. microsc. Soc. 66: 69–76.

Shikata, E. & K. Maramorosch, 1965. Plant tumour virus in arthropod host: microcrystal formation. – Nature 208: 507–508.

Shikata, E. & K. Maramorosch, 1966. Electron microscopy of pea enation mosaic virus in plant cell nuclei. – Virology 30: 439–454.

Shikata, E. & Maramorosch, 1967. Electron microscopy of wound tumor virus assembly sites in insect vectors and plants. – Virology 32: 363–377.

Shikata, E., K. Maramorosch & R. R. Granados, 1966. Electron microscopy of pea enation mosaic virus in plants and aphid vectors. – Virology 29: 426–436.

Shikata, E., S. W. Orenski, H. Hirumi, J. Mitsuhashi & K. Maramorosch, 1964. Electron micrographs of wound-tumor virus in an animal host and in a plant tumor. – Virology 23: 441–444.

Shimomura, T., 1971. Necrosis and localization of infection in local lesion hosts. – Phytopath. Z. 70: 185–196.

Singh, M. & A. C. Hildebrand, 1966. Cinephotomicrography of tobacco mosaic virus inclusions in cultured tobacco cells. – Phytopathology 56: 901.

Slogteren, D. H. M. van, 1958. Ratelvirus als oorzaak van ziekten in bloembolgewassen en de mogelijkheden de infectie door middel van grondontsmetting te bestrijden. (With a summary: Rattle virus as the cause of diseases in flower bulbs and the possibilities of controlling infection by means of soil desinfectants). – Tijdschr. PlZiekt. 64: 452–462.

Slogteren, E. van & M. P. de Bruyn Ouboter, 1941. Onderzoekingen over virusziekten in bloembolgewassen II. Tulpen I. (With a summary). – Meded. LandbHoogesch. Wageningen 45.

Smith, K. M., 1935. New virus diseases of the tomato. – Jl R. hort. Soc. 60: 448–451.

Smith, K. M., 1951. A latent virus in sugar beets and mangolds. – Nature (Lond.) 167: 1061.

Smith, K. M., 1958. Virus inclusions in plant cells. – Protoplasmatologia IV 4a.

Smith, R. E. & H. E. Thomas, 1951. Prune diamond canker. – U.S.D.A. Agriculture Handbook 10: 175–176.

Solberg, R. A. & J. G. Bald, 1962. Virus invasion and multiplication during leaf histogenesis. – Virology 17: 359–361.

Sommereyns, G., 1967. Les virus des végétaux, leurs propriétés et leur identification. – 2ième éd. J. Duculot S.A. Gembloux.

Sprau, F., 1955. Pathologische Gewebeveränderungen durch das Blattrollvirus bei der Kartoffel und ihr färbetechnischer Nachweis. – Ber. dt. bot. Ges. 68: 239–246.

Steere, R. L., 1952. Virus increment curves obtained from counts of particles in clarified plant juice. – Am. J. Bot. 39: 211–220.

Steere, R. L. & R. C. Williams, 1953. Identification of cristalline inclusion bodies extracted intact from plant cells infected with tobacco mosaic virus. – Am. J. Bot. 40: 81–84.

Štefanac, Z. & D. Miličić, 1965. Zelleinschlüsse des Kohlrübenmosaikvirus. – Phytopath. Z. 52: 349–362.

Stoll, K., 1952. Über die Symptome der Pfeffingerkrankheit der Kirschbaume. – Phytopath. Z. 18: 293–306.

Stubbs, L. L., 1947. A destructive vascular wilt virus disease of broad bean (*Vicia faba* L.) in Victoria. – J. Dep. Agric. Vict. 46: 323–332.

Su, M. T., 1933. Rep. Mycol. Burma. – In: Rev. appl. Mycol. 13, 1934, 78.
Swingle, R. U., 1938. A phloem necrosis of elm. – Phytopathology 28: 757–759.
Swingle, R. U. & T. W. Bretz, 1950. Zonate canker, a virus disease of American elm. – Phytopathology 40: 1018–1022.
Sylvester, E. S., 1948. The yellow-net virus disease of sugar beets. – Phytopathology 38: 429–439.
Symposium, 1958. Symposium latency and masking in viral and Rickettsial infection. – Proc. Conf. Univ. Wisc. Medical School Sept. 4–6, 1957.
Takahashi, T., 1972. Studies on viral pathogenesis in plant hosts. II. Changes in developmental morphology of tobacco plants infected systemically with tobacco mosaic virus. – Phytopath. Z. 74: 37–47.
Takahashi, T., 1975. Studies on viral pathogenesis in plant hosts. VIII Systemic invasion and localization of infection on 'Samsun NN' tobacco plants resulting from tobacco mosaic virus infection. – Phytopath. Z. 84: 75–87.
Takahashi, W. N. & M. Ishii, 1953. A macromolecular protein associated with tobacco mosaic virus infection: its isolation and properties. – Am. J. Bot. 40: 85–90.
Tepfer, S. S. & M. Chessin, 1959. Effects of tobacco mosaic virus on early leaf development in tobacco. – Am. J. Bot. 46: 496–509.
Thomas, H. R. & W. J. Zaumayer, 1950. Red node, a virus disease of beans. – Phytopathology 40: 832–846.
Thomas, K. M. & C. S. Krishnaswami, 1939. Little leaf, a transmissible disease of brinjal. – Proc. Indian Acad. Sci. B 10: 201–212.
Thresh, J. M., 1967. Increased susceptibility of black-currant bushes to the gall-mite vector (*Phytoptus ribis* Nal.) following infection with reversion virus. – Ann. appl. Biol. 60: 455–467.
Tinsley, T. W., 1959. Pea leaf roll, a new virus disease of legumes in England. – Pl. Path. 8: 17–18.
Tjallingii, F., 1952. Onderzoekingen over de mozaïekziekte van de augurk (*Cucumis sativus* L.) (With summary: Investigations on the mosaic disease of gherkin, *Cucumis sativus* L.). – Meded. Inst. plziektenk. Onderz. Wageningen, 47.
Todd, J. M., 1958. Witches' broom of potatoes and similar diseases in Great Britain. – Proc. Sci. Conf. Stolbur and similar diseases causing sterility, Smolenice 1956: 77–101.
Tompkins, C. M., 1939. Two mosaic diseases of annual stock. – J. Agric. Res. 58: 63–77.
Toxopeus, H. J., 1936. Stock-scion incompatibility in citrus and its cause. – J. Pomol. 14: 360–364.
Trotter, A., 1954. Virosi delle piante e cecidologia. – Marcellia 30: 10–15.

Ullrich, J. & L. Quantz, 1964. Morphologische und anatomische Untersuchungen über pathologische Veränderungen durch das Scharfe Adernmosaik (pea enation mosaic virus) an *Vicia faba* und *Pisum satisvum* und durch eine Virose an *Trifolium repens*. – Phytopath. Z. 51: 1–18.

Uschdraweit, H. A., 1967. Symptomatologie: a. Äussere Symptome. In: M. Klinkowski, ed.: Pflanzliche Virologie. I. Einführung in die allgemeinen Probleme. 2e Aufl.: 21–55. Akademie Verl. Berlin.

Ushiyama, R. & R. E. F. Matthews, 1970. The significance of chloroplast abnormalities associated with infection by turnip yellow mosaic virus. – Virology 42: 293–303.

Valenta, V., M. Musil & S. Mišiga, 1961. Investigations on European yellows-type viruses. I. The stolbur virus. – Phytopath. Z. 42: 1–38.

Valleau, W. D., 1932. Seed transmission and sterility studies of two strains of tobacco ring spot. – Bull. Ky agric. Exp. Stn. 327: 43–80.

Venekamp, J. H., 1957. Onderzoek naar biochemische verschillen tussen viruszieke en gezonde planten. – Jaarversl. Inst. PlZiektenk. Onderzoek 1956: 87–90.

Venekamp, J. H. & W. H. M. Mosch, 1963. Chromatographic studies on plant viruses. I. The isolation of potato virus X by means of various systems of adsorption chromatography. – Virology 19: 316–321.

Vries, H. de, 1896. Een epidemie van vergroeningen (Avec un résumé français: Une épidémie de virescences). – Bot. Jaarb. 8: 66–91.

Wache, H. W., 1966. Einfluss von Temperatur und Luftfeuchtigkeit auf Virussymptome. – Mber. dt. Akad. Wiss. Berl. 8: 754–757.

Walkey, D. G. A. & M. J. W. Webb, 1970. Tubular inclusion bodies in plants infected with viruses of the NEPO type. – J. gen. Virol. 7: 159–166.

Walkinshaw, C. H. & R. H. Larson, 1959. Corky ringspot of potato, a soil-borne virus disease. – Res. Bull agric. Exp. Stn Univ. Wis. 217.

Wallace, J. M., 1944. Acquired immunity from curly top in tobacco and tomato. – J. agric. Res. 69: 187–214.

Wallace, J. M. & R. J. Drake, 1961. Induction of woody galls by wounding of citrus infected with vein enation virus. – Pl. Dis. Reptr 45: 682–686.

Want, J. P. H. van der, 1948. Het stippelstreep van de boon (*Phaseolus vulgaris*), een ziekte veroorzaakt door een virus dat in de grond overblijft (With a summary: Stipple-streak of the bean (*Phaseolus vulgaris*) a disease caused by a soil-borne virus). – Tijdschr. PlZiekt. 54: 85–90.

Want, J. P. H. van der, 1969. The virion, an outdated conception? – Neth. J. Pl. Path. 75: 96–99.

Want, J. P. H. van der & L. Bos, 1959. Geelnervigheid, een virusziekte van luzerne. (With a summary: Vein yellowing a virus disease of lucerne.) – Tijdschr. PlZiekt. 65: 73–78.

Warmke, H. E., 1969. A reinterpretation of amorphous inclusions in the aucuba strain of tobacco mosaic virus. – Virology 39: 695–704.

Warmke, H. E. & J. R. Edwardson, 1966. Electron microscopy of crystalline inclusions of tobacco mosaic virus in leaf tissue. – Virology 30: 45–57.
Wasuwat, S. L. & J. C. Walker, 1961a. Inheritance of resistance in cucumber to cucumber mosaic virus. – Phytopathology 51: 423–428.
Wasuwat, S. L. & J. C. Walker, 1961b. Relative concentration of cucumber mosaic virus in a resistant and a susceptible cucumber. variety. – Phytopathology 51: 614–616.
Watson, M. A. & E. P. Serjeant, 1962. Carrot motley dwarf. – Rep. Rothamsted exp. Stn 1961: 106–107.
Watson, M. A. & D. J. Watson, 1953. Comparative physiological studies on the growth of field crops. III The effect of infection with beet yellows and beet mosaic viruses on the growth and yield of the sugar-oat root crop. – Ann. appl. Biol. 40: 1–37.
Watson, R. D. & J. W. Guthrie, 1964. Virus-fungus interrelationships in a root rot complex in red clover. – Pl. Dis. Reprt 48: 723–727.
Webb, R. E. & Schultz, E. S., 1958. Possible relation between haywire of potato and big bud of tomato. – Pl. Dis. Reptr. 42: 44–47.
Wehrmeyer, W., 1960. Die Bildung 'intracellular' Stäbe nach Virus-infection. – Naturwissenschaften 47: 236–237.
Weintraub, M. & H. W. J. Ragetli, 1966. Fine structure of inclusions and organelles in *Vicia faba* infected with bean yellow mosaic virus. – Virology 28: 290–302.
Weintraub, M. & H. W. J. Ragetli, 1968. Intracellular characterization of bean yellow mosaic virus-induced inclusions by differential enzyme digestion. – J. Cell Biol. 38: 316–328.
Weintraub, M. & H. W. J. Ragetli, 1970. Electron microscopy of the bean and cowpea strains of southern bean mosaic virus within leaf cells. – J. ultrastruct. Res. 32: 167–189.
Weintraub, M. & H. W. J. Ragetli, 1971. A mitochondrial disease of leaf cells infected with an apple virus. – J. ultrastruct. Res. 36: 669–693.
Weintraub, M., H. W. J. Ragetli & E. Leung, 1976. Elongated virus particles in plasmodesmata. – J. ultrastruct. Res. 56: 351–364.
Wenzl, H., 1964. Die Welkekrankheit der Kartoffel. – Pflanzenschutzberichte 31: 161–178.
Whipple, O. C. & J. C. Walker, 1941. Strains of cucumber mosaic virus pathogenic on bean and pea. – J. agric. Res. 62: 27–60.
Williams, R. C. & R. W. G. Wyckoff, 1945. Electron shadow micrography of virus particles. Proc. Soc. exp. – Biol. Med. 58: 265–270.
Wills, A. B., 1962. Genetical aspects of strawberry June yellows. – Heredity 17: 361–372.
Wilson, J. H., 1955. Effects of nutrition and light intensity on symptoms of leaf-roll virus infection in the potato plant. – Ann. appl. Biol. 43: 273–287.
Windsor, I. M. & L. M. Black, 1972. Clover club-leaf; a possible rickettsial disease of plants. – Phytopathology 62: 1112.

Wingard, S. A., 1928. Hosts and symptoms of ring spot, a virus disease of plants. – J. agric. Res. 37: 127–153.
Wolffgang, H., 1970a. Der Einfluss der Temperatur auf die Virusvermehrung und auf die Symptomausbildung infizierter Pflanzen. – Arch. Pflanzensch. 6: 353–382.
Wolffgang, H., 1970b. Einfluss von Umweltbedingungen auf Virusvermehrung und Symptombildung. 2. Einfluss der Temperatur auf Grösse, Wachstum und Zeitpunkt des Auftretens von Lokalläsionen auf *Nicotiana glutinosa* L. und *Phaseolus vulgaris* L. var. Pinto durch TMV. – Acta Phytopathol. Acad. Sci. Hung. 5: 177–180.
Yamaguchi, A., 1964. Current season symptoms in tulip flowers by the tulip breaking virus. – Phytopath. Z. 49: 355–361.
Yamaguchi, A. & T. Hirai, 1967. Symptom expression and virus multiplication in tulip petals. – Phytopathology 57: 91–92.
Yarwood, C. E., 1959. Predisposition. In: J. G. Horsfall & A. E. Dimond, eds. – Plant Pathology I: 521–562. Acad. Press. New York, London.
Yarwood, C. E., E. C. Resconich, P. A. Ark, D. E. Schlegel & K. M. Smith, 1961. So-called beet latent virus is a bacterium. – Pl. Dis. Rept. 45: 85–89.
Zaumeyer, W. J. & H. R. Thomas, 1948. Pod mottle, a virus disease of beans. – J. agric. Res. 77: 81–96.
Zech, H., 1952. Untersuchungen über den Infektionsvorgang und die Wanderung des Tabaksmosaikvirus im Pflanzenkörper. – Planta 40: 461–514.
Zeller, S. M., J. R. Kienholz & J. A. Milbrath, 1951. Black canker. – U.S. D. A. Agriculture Handbook 10: 137–138.
Zycha, H., 1955. Definition von Rindenbrand und Krebs bei Waldbäumen. – Meded. LandbHoogesch. OpzoekStns Gent 20: 411–418.

Index of names of symptoms

The terms considered the most appropriate are always mentioned first. It is suggested that the names in parenthesis be eliminated from symptom description. However they are given here because they occur in older publications or are still used occasionally. Their listing here also makes the key to the text a complete one. Names in quotation marks are used to indicate diseases rather than symptoms.

Abbreviations

d = Dutch; **g** = German; **f** = French; **i** = Italian; **s** = Spanish.

I. English index with Dutch, German, French, Italian and Spanish equivalents

The page numbers in this list refer to those pages only where the terms have been italicized and defined. For a complete reference to text and figures see subject index.

8 asteroid spotting 64, 69
d stervormige vlekking
g Sternfleckung
f taches étoilées
i macchiettatura asteroide
s manchas en estrella

9 atrophy 52
d atrofie
g Atrophie
f atrophie
i atrofia
s atrofia

10 (aucuba mosaic) 62
d (aucubamozaïek, aucubabont)
g (Aucubamosaik)
f (mosaïque aucuba)
i (mosaico aucuba)
s (mosaico aucuba)

11 bark necrosis 80
d bastnecrose
g Rindenbrand
f nécrose corticale
i necrosi corticale
s necrosis de la corteza, necrosis cortical

12 bark scaling 86
d schubbigheid
g Schuppigkeit
f desquamation corticale
i desquamazione della corteccia
s descarnación cortical

13 blackening 57, 67, 69
d zwartkleuring
g Schwarzfärbung
f noircissement
i annerimento
s ennegrecimiento

14 blanching, bleaching 56, 69
d verbleking, witkleuring
g Bleichung, ('Bleiche')
f blanchiment
i sbiancatura, imbianchimento
s decoloración

15 blastomania, see witches' broom growth 106

16 bleaching, see blanching

17 blotching, see diffuse mottling

18 branch flattening 88, 99
d takafplatting
g Flachästigkeit, Zweigabplattung
f aplatissement des branches
i appiattimento dei rami
s aplanamiento de las ramas

19 bronzing 57, 67, 69, 77
d bronskleuring
g Bronzefärbung
f bronzage
i bronzatura
s bronceado

20 browning 57, 67, 69
d verbruining
g Braunfärbung, ('Bräune')
f brunissement
i imbrunimento
s amarronamiento

21 (calico) 62
d (calico)
g (Kaliko)
f (calicot)
i (calico)
s (calico)

22 callose formation, (excessive) – 49

d (overmatige) callosevorming
g (übermässige) Kallosebildung
f formation (excessive) de callose
i formazione (eccessiva) di callosi
s formación (excesiva) de callosa

23 canker 80
d kanker
g Krebs
f chancre
i cancro (vegetale)
s chancro

24 cankerous tumour, cf. canker and tumour 82, 88, 98

25 chlorosis 58, 69
d chlorose
g Chlorose
f chlorose
i clorosi
s clorosis

26 chronic phase 27
d chronische fase
g chronische Phase
f phase chronique
i fase cronica
s fase crónica

27 cladomania, see witches' broom growth 106

28 contamination 16
d besmetting
g Kontaminierung, Verunreinigung
f contamination
i contaminazione
s contaminación

29 cork formation 40, 85
d kurkvorming
g Korkbildung
f subérisation
i formazione di sughero, suberosi
s suberización

30 correlation disturbance 52, 113
d correlatiestoring
g Korrelationsstörung
f perturbation de la corrélation hormonale
i alterazione dell'equilibrio ormonale
s trastorno hormonal del crecimiento

31 crinkling, crumpling, wrinkling 88, 89, 100
d kreuk(el)ing, rimpeling, (krinkeling)
g Kräuselung, leichte Kräuselung, Verbeulung
f frisolée faible, gaufrage
i arricciamento leggero
s arrugamiento ligero

32 curling, leaf curling 88, 89, 100
d krulling, bladkrulling
g Kräuselung, schwere Kräuselung
f frisolée grave
i arricciamento grave
s rizado de la hoja

33 dappling, see flecking

34 decline 25

d aftakeling
g Verfall
f dépérissement
i deperimento, moria

s decaimiento

35 defective lignification 49
d onvolledige verhouting
g unvollständige Verholzung
f lignification défectueuse
i lignificazione incompleta
s lignificación incompleta

36 defoliation, see leaf dropping

37 deformation, see malformation

38 degeneration 27, 53

d degeneratie
g Degeneration, Abbau
f dégénérescence
i degenerazione
s degeneración

39 desiccation 73
d verdroging, (uitdroging)
g Vertrocknung, (Austrocknung)
f dessèchement
i disseccamento
s desecación

40 diffuse mottling, blotching, splotching 64, 69, 84
d vlekkerigheid
g Scheckung
f marbrure diffuse
i maculatura diffusa
s moteado difuso

41 disease 17, 34
d ziekte
g Krankheit
f maladie
i malattia
s enfermedad

42 disease symptom, see symptom

43 distortion 88, 89
d verwringing, distorsie
g Verdrehung, Verkrümmung
f distorsion
i distorsione
s distorsión

44 disturbed periodicity 110
d verstoorde periodiciteit
g gestörte Periodizität
f altération de la périodicité
i alterazione del periodismo
s alteración de la periodicidad

45 dotting, see speck(l)ing

46 double nodes 88, 103
d dubbele knopen
g Doppelknoten, (Kurzknoten)
f doubles noeuds
i nodi doppi
s doble nudos

47 dwarfing, stunting 54, 88
d dwerggroei
g Zwergwuchs, Kümmerwuchs, Nanismus, Stauchung, Verzwergung
f nanisme, rabougrissement
i nanismo, rachitismo, arresto di sviluppo
s enanismo

48 edge chlorosis 58, 69
d randchlorose, bladrandchlorose
g Randchlorose, Blattrand-chlorose
f chlorose du bord de la feuille
i clorosi marginale
s clorosis marginal

49 edge yellowing 58, 69
d randvergeling, bladrand-vergeling
g Randvergilbung, Blattrand-vergilbung
f jaunissement du bord de la feuille
i ingiallimento marginale (della foglia)
s amarilleamiento marginal

50 enation 93, 102
d enatie, uitwas
g Enation, Auswuchs
f énation
i enazione, escrescenza
s excrecencia, (enation)

51 epinasty 88, 102
d epinastie
g Epinastie
f épinastie
i epinastia
s epinastia

52 etching 67, 69, 73, 77
d etsing
g Ätzung, Ätzmuster
f gravure
i 'etching'
s grabado

53 fanleaf formation 88, 100
d waaierbladvorming
g Fächerblattbildung
f formation de la feuille en éventail
i formazione di foglie a ventaglio
s formación en abanice

54 flecking, dappling, freckling 64, 69
d vlekking, gevlektheid
g Fleckung, Fleckenbildung
f tacheture
i maculatura
s manchado, (flecking)

55 flower colour breaking 70
d bloemkleurbreking
g Blütenfarbenbrechung, (Buntstreifigkeit der Blumen)
f panachure de la fleur
i rottura del colore, variegatura dei fiori
s rotura del color de las flores

56 freckling, see flecking

57 graft incompatibility 115
d onverenigbaarheid bij enting
g Pfropfunverträglichkeit
f incompatibilité de greffage
i incompatibilità di innesto
s incompatibilidad de injerto

58 greening, virescence 71, 108
d vergroening, virescentie
g Vergrünung, Vireszenz
f virescence
i inverdimento, virescenza
s reverdimiento

59 growth reduction 40, 54, 88
d groeiremming
g Wachstumshemmung, Wuchs-hemmung
f réduction de (la) croissance
i riduzione di sviluppo
s reducción del crecimiento

60 gummosis 50
d gomvorming
g Gummibildung, Gummosis

f gommose
i gommosi
s gomosis

61 hair sprouting, see spindle malformation

62 histoid malformation, see sprouting

63 hyperplasia 50
d hyperplasie
g Hyperplasie
f hyperplasie
i iperplasia
s hiperplasia

64 hypersensitivity 18, 21, 28, 75
d overgevoeligheid
g Überempfindlichkeit, Hypersensibilität
f hypersensibilité
i ipersensibilità
s hipersensibilidad

65 hypertrophy 50, 88
d hypertrofie
g Hypertrophie
f hypertrophie
i ipertrofia
s hipertrofia

66 hypoplasia 52
d hypoplasie
g Hypoplasie
f hypoplasie
i ipoplasia
s hipoplasia

67 immunity 17
d immuniteit, onvatbaarheid
g Immunität
f immunité
i immunità
s immunidad

68 inapparency, inapparent infection 17, 30, 34
d inapparentie, symptoomloosheid, inapparente infectie, symptoomloze infectie
g Inapparenz, inapparente Infektion
f infection inapparente
i infezione non manifesta
s infección inaparente

69 inapparent infection, see inapparency

70 inclusion bodies 122–138
d celinsluitsels
g Einschlusskörper
f inclusions
i inclusioni cellulari
s inclusiones intracelulares

71 incubation period 17, 32
d incubatieperiode, incubatietijd
g Inkubationsperiode, Inkubationszeit
f période d'incubation
i periodo d'incubazione
s periodo de incubación

72 infection 16
d infectie
g Infektion
f infection
i infezione
s infección

73 inoculation 16
d inoculatie, inenting
g Impfung, (Beimpfung eines

Blattes, Verimpfung eines infektiösen Saftes)
f inoculation
i inoculazione
s inoculación

74 internode shortening 88, 103
d internodienverkorting, stengellidverkorting
g Internodienverkürzung
f raccourcissement des entrenoeuds
i accorciamento degli internodi
s acortamiento de los entrenudos

75 interveinal mosaic 62, 69
d tussennervig mozaïek
g Zwischennervenmosaik
f mosaïque internervaire
i mosaico internervale
s mosaico intervenal

76 (lacing), see shoe stringing

77 latency 32
d latentie
g Latenz
f latence, état latent
i latenza
s latencia

78 leaf abscission, see leaf dropping

79 leaf casting, see leaf dropping

80 leaf curling, see curling

81 leaf dropping, leaf abscission, leaf casting, leaf fall, defoliation 113
d bladval
g Blattfall
f abscission des feuilles, chute des feuilles
i filloptosi, caduta di foglie
s defoliación

82 leaf fall, see leaf dropping

83 leaf narrowing 87, 88, 100
d smalbladigheid
g Schmalblättrigkeit, Blattverschmälerung
f rétrécissement du limbe
i restringimento della foglia, stenofillia
s estenofilia, estrechamiento de la hoja

84 leafrolling 88, 89, 102
d bladrolling
g Blattrollen, ('Blattroll')
f enroulement foliaire
i accartocciamento, arrotolamento della foglia
s enrollamiento de la hoja

85 lesion 20, 23
d lesie
g Läsion
f lésion
i lesione
s lesión

86 line pattern 23, 69
d figuurbont
g Linien- und Kurvenmuster
f arabesques
i lineolatura
s arabescos

87 local lesion 20, 75
d lokale lesie
g Lokalläsion, (Einzelherd), (Lokalherd)
f lésion locale
i lesione locale

s lesión local

88 local symptoms 20
d lokale symptomen
g Lokalsymptome
f symptômes localisés, symptômes locaux
i sintomi locali
s síntomas locales

89 male sterility 114
d mannelijke steriliteit
g männliche Sterilität
f stérilité mâle
i sterilità maschile
s esterilidad masculina

90 malformation, deformation 87, 88
d misvorming
g Missbildung, Bildungsabweichung, Deformation
f malformation
i malformazione, deformazione
s deformación

91 masking 33
d maskering
g Maskierung
f masquage
i mascheramento
s enmascaramiento

92 mosaic 61, 69
d mozaïek
g Mosaik
f mosaïque
i mosaico
s mosaico

93 (mottle), see mottling

94 mottling (mottle), cf. diffuse mottling 64, 69
d gevlektheid
g Fleckung
f marbrure
i mosaicatura (alquanto sfumata), maculatura
s moteado

95 necrosis 40, 52, 74
d necrose
g Nekrose
f nécrose
i necrosi
s necrosis

96 negative geotropism, intensification of – 111
d versterking der negatieve geotropie
g Verstärkung der negativen Geotropie
f intensification du géotropisme négatif
i accentuazione del geotropismo negativo
s intensificación del geotropismo negativo

97 oak-leaf pattern 23, 66, 69
d eikebladpatroon
g Eichenblattmuster
f taches digitées
i maculatura a foglia di quercia
s manchas en hoja de roble

98 oil fleck 65, 69
d olievlek
g Ölfleck
f tache huileuse
i macchia d'olio
s mancha aceitosa

99 organoid malformation, see malformation

100 pathogenesis 12, 16, 41
d pathogenese
g Pathogenese, Krankheitsentstehung
f pathogénèse
i patogenesi
s patogenesis

101 phloem degeneration, cf. degeneration 53
d floëemdegeneratie, zeefvatdegeneratie
g Phloemdegeneration
f dégénérescence du liber
i degenerazione del floema
s degeneración del floema

102 phloem necrosis 79, 84
d floëemnecrose, zeefvatnecrose
g Phloemnekrose
f nécrose du liber
i necrosi del floema
s necrosis del floema

103 phyllody 88, 108
d fyllodie
g Phyllodie
f phyllodie
i fillodia
s filodia

104 pollen abortion, pollen sterility 114
d stuifmeelabortie, pollensteriliteit
g Pollenabortion, Pollensterilität
f avortement du pollen, stérilité du pollen
i aborto del polline, sterilità del polline
s esterilidad del polen

105 pollen sterility, see pollen abortion

106 predisposition 18, 117
d predispositie
g Prädisposition
f prédisposition
i predisposizione
s predisposición

107 primary symptoms 22
d primaire symptomen
g Primärsymptome, (Initialsymptome)
f symptômes primaires
i sintomi primari
s síntomas primarios

108 proliferation 50, 91
d woekering, (proliferatie)
g Wucherung
f prolifération
i proliferazione
s proliferación

109 prolification 110
d prolificatie, (bloemdoorgroei)
g Prolifikation, (Blütendurchwuchs)
f prolification
i prolificazione, diafisi
s prolificación

110 purpling 57, 66, 69
d paarskleuring
g Purpurfärbung
f coloration pourpre
i colorazione purpurea
s purpurescencia

111 recovery 27, 34
d herstel
g Erholung
f rétablissement

i ripresa
s reponerse

112 recurrency 33
d herhaling
g Wiedererscheinung, Wieder-auftreten
f récurrence
i ricomparsa (dei sintomi)
s recurrencia

113 reddening 57, 66, 69
d roodkleuring
g Rotfärbung, ('Röte')
f rougissement
i arrossamento
s enrojecimiento

114 reduction of seed dormancy 110
d kiemrustverkorting
g Abkürzung der Keimruhe
f réduction de la durée du repos germinatif des graines
i riduzione della quiescenza dei semi
s reducción del período de reposo germinativo de los semillas

115 reduction of vitality 27
d achteruitgang van de vitaliteit
g Vitalitätsminderung, Vitalitätsrückgang
f réduction de vitalité
i riduzione di vitalità
s reducción de la vitalidad

116 resistance 18, 33
d resistentie
g Resistenz, Widerstands-fähigkeit
f résistance
i resistenza
s resistencia

117 ring formation 69, 76, 82
d kringerigheid, kringvorming
g Ringbildung, Kringelbildung, Pfropfenbildung (bei Kartoffel-knollen)
f formation de cernes, formation d'anneaux
i anulatura
s formación de anillos

118 ringspotting 64, 69, 76
d kringvlekvorming
g Ringfleckung, Ringflecken-bildung, Ringfleckigkeit
f formation de taches annulaires
i maculatura anulare
s manchas en anillo

119 rosetting 88, 103
d rozetvorming
g Rosettenbildung
f formation de rosettes
i sviluppo a rosetta
s formación de rosetas

120 rotting 75, 118
d rotting
g 'Fäule'
f pourriture
i marciume, marcescenza
s podredumbre

121 'rough skin' 86
d ruwschilligheid
g Rauhschaligkeit
f peau rugueuse
i buccia ruvida
s piel rugosa

122 'rubbery wood' 49
d 'rubberhout'

g 'Gummiholz'
f 'bois souple', 'bois caoutchouc'
i 'legno di gomma, 'legno di caucciù'
s madera de consistencia gomosa

123 rugosity 88, 100
d diepnervigheid
g Rauhblättrigkeit, Tiefadrigkeit
f rugosité, gaufrage
i rugosità
s rugosidad

124 secondary symptoms 22
d secondaire symptomen
g Folgesymptome, Sekundärsymptome
f symptômes secondaires
i sintomi secondari
s sintomas secundarios

125 seed abortion 85, 114
d zaadabortie
g Samenabortion
f avortement des graines
i aborto dei semi
s detención del crecimiento de la semilla

126 sensitivity 17, 32
d gevoeligheid
g Empfindlichkeit, Sensitivität
f sensibilité
i sensibilità, sensitività
s sensibilidad

127 shock phase, see acute phase

128 shoe stringing, (lacing) 88, 100
d draadbladigheid, naaldbladigheid
g Fadenblättrigkeit
f formation de feuilles filiformes
i formazione di foglie filiformi
s formación de las hojas filiformes

129 shoot swelling 88, 98
d scheutzwelling, takzwelling
g Triebschwellung
f renflement des branches
i ingrossamento dei germogli (dei rami), rigonfiamento dei germogli (dei rami)
s hinchazón del tallo (y de la rama)

130 shot-hole 77
d hagelschot
g Schrotschusseffekt
f criblure
i perforazioni (fogliari)
s lesiones en agujero

131 speck(l)ing, stippling, spotting, dotting 64, 69, 75
d spikkeling, stippeling
g Sprenkelung, Tüpfelung
f moucheture
i punteggiatura, puntulatura, macchiettatura
s mancha

132 spindle sprouting, hair sprouting 112
d draadspruitigheid
g Fadenkeimigkeit
f filosité
i formazione di germogli filanti
s brotes filiformes

133 splotching, see diffuse mottling

134 spotting, see speckling

135 star cracking 86

d vorming van stervormige barsten
g Sternrissigkeit
f craquelure en étoile
i spaccatura stellata, spaccatura stellare
s rotura en estrella

136 starch accumulation 42
d zetmeelophoping
g Stärkestauung, Stärkeschoppung
f accumulation d'amidon
i accumulo di amido
s acumulación de almidón

137 starch lesion 42, 49
d zetmeellesie
g Stärkeläsion
f lésion d'amidon
i lesione di amido
s lesión de almidón

138 stem grooving 88, 99
d vorming van houtgroeven
g Stammfürchung
f bois sillonné
i grandi depressioni (nel fusto)
s asurcamiento del tallo

139 stem pitting 88, 98
d vorming van houtputjes
g Stammnarbung
f bois strié
i piccole depressioni nel fusto
s formación de pocillos en el tallo

140 sterility 88, 110
d steriliteit
g Sterilität
f stérilité
i sterilità
s esterilidad

141 stippling, see speck(l)ing

142 streaking, striping 61, 69, 70, 80
d streping
g Strichelung, Streifung
f striure
i striatura
s estriado

143 striping, see streaking

144 stunting, see dwarfing

145 susceptibility 17, 18, 117
d vatbaarheid
g Anfälligkeit, Empfänglichkeit
f susceptibilité, réceptivité
i recettività, suscettibilità
s susceptibilidad, receptividad

146 symptom, disease symptom 10
d symptoom, ziektesymptoom, ziekteverschijnsel
g Symptom, Krankheitssymptom, Krankheitserscheinung
f symptôme (de maladie)
i sintomo (della malattia)
s síntoma (de enfermedad)

147 syndrome 17
d syndroom, ziektebeeld
g Syndrom, Krankheitsbild
f syndrome
i sindrome
s sindrome

148 systemic symptoms 22
d systemische symptomen, (spreidsymptomen)
g systemische Symptome

f symptômes généralisés
i sintomi sistemici
s síntomas generalizados

149 (teratomata) 88, 99
d (teratomata)
g (Teratomata)
f (tératomes)
i (forme teratologiche)
s (teratoma)

150 tissue degeneration, cf. also degeneration 53
d weefseldegeneratie
g Gewebedegeneration
f dégénérescence des tissus
i degenerazione dei tessuti
s degeneración del tejido

151 tolerance 17, 32
d tolerantie
g Toleranz
f tolérance
i tolleranza
s tolerancia

152 top necrosis 78
d topnecrose, topsterfte, (acronecrose)
g Spitzennekrose, Akronekrose
f nécrose apicale
i necrosi apicale
s necrosis apical

153 trabeculae 50
d trabeculae
g intrazellulare Stäbe
f cordons endocellulaires, trabécules
i trabecule, cordoni endocellulari
s trabecula

154 tumour (Am. tumor) 88, 95
d tumor, gezwel
g Tumor, Wucherung
f tumeur
i tumore
s tumor

155 tumorous canker, see canker and tumour 82, 88, 98

156 tylose formation, (excessive) –, tylosis 50
d (overmatige) thyllenvorming
g (übermässige) Thyllenbildung
f formation (excessive) de thylles, thyllose
i tillosi, (eccessiva) formazione di tilli
s formación (excesiva) de tilosa

157 variegation 60, 69
d bontheid
g Buntblättrigkeit, (Variegation)
f variégation
i variegatura
s variegación

158 vein(al) necrosis 22, 77
d nerfnecrose
g Adernnekrose, ('Rippenbräune')
f nécrose des nervures
i necrosi delle nervature
s necrosis de la nervadura

159 vein banding 62, 69
d nerfbandmozaïek
g Adernbänderung, Adernbandmosaik
f liseré des nervures
i bandatura nervale
s formación de bandas paralela a la nervadura

160 vein chlorosis 22, 58, 69
d nerfchlorose
g Adernchlorose
f chlorose des nervures
i clorosi delle nervature, clorosi nervale
s chlorosis de la nervadura

161 vein clearing 58, 69
d nerfglazigheid
g Adernaufhellung
f éclaircissement des nervures
i schiarimento delle nervature
s aclaramiento de la nervadura

162 vein mosaic 62, 69
d nerfmozaïek
g Adernmosaik
f mosaïque des nervures
i mosaico nervale
s mosaico de las venas

163 vein yellowing 58, 69
d nerfvergeling, (geelnervigheid)
g Adernvergilbung
f jaunissement des nervures
i ingiallimento delle nervature
s amarilleamiento de la nervadura

164 virescence, see greening

165 water deficiency, water shortage 40, 52, 73
d watertekort, watergebrek
g Wasserdefizit
f déficience en eau, manque d'eau
i deficienza idrica, carenza idrica
s falta de agua, insuficiencia hídrica, deficiencia hídrica

166 water shortage, see water deficiency

167 wilting 73, 77
d verwelking
g Verwelken, Welken, ('Welke')
f flétrissement
i appassimento
s marchitamiento

168 witches' broom formation 106
d heksenbezemvorming
g Hexenbesenbildung
f formation de balais de sorcière
i formazione di scopazzi
s formación en escoba de bruja

169 witches' broom growth, blastomania, cladomania 106, 110
d heksenbezemgroei, blastomanie, cladomanie
g Hexenbesenwuchs, Blastomanie, Kladomanie, Triebsucht
f blastomanie, croissance en balais de sorcière
i blastomania, cladomania, sviluppo di scopazzi
s crecimiento en escoba de bruja, blastomania

170 witches' broom phenomena 104, 110
d heksenbezemverschijnselen
g Hexenbesenerscheinungen
f phénomènes des balais de sorcière
i fenomeni di scopazzatura
s fenómeno de escoba de bruja

171 withering 73, 77
d verdorring

g Verdorren, ('Dürre')
f dessèchement
i avvizzimento
s desecamiento

172 yellowing 56, 57, 69
d vergeling
g Vergilbung
f jaunissement
i ingiallimento
s amarilleamiento, clorosis amarilla

173 yield reduction 55
d opbrengstdaling, oogstreductie
g Ertragsminderung
f diminution du rendement
i riduzione di raccolto, diminuzione di prodotto
s disminución del rendimiento

174 zigzag growth 88, 103
d zigzag-groei
g Zickzackwuchs
f croissance en zigzag
i sviluppo a zig zag
s crecimiento en zig zag

II. Dutch index of names of symptoms

N.B. The numbers correspond to those of the English index with equivalents in the other languages. The English list serves as an index to the text of this book.

III. German index of names of symptoms

N.B. The numbers correspond to those of the English index with equivalents in the other languages. The English list serves as an index to the text of this book.

IV. French index of names of symptoms

N.B. The numbers correspond to those of the English index with equivalents in the other languages. The English list serves as an index to the text of this book.

V. Italian index of names of symptoms

N.B. The numbers correspond to those of the English index with equivalents in the other languages. The English list serves an as index to the text of this book.

VI. Spanish index of names of symptoms

N.B. The numbers correspond to those of the English index with equivalents in the other languages. The English list serves as an index to the text of this book.

Subject index

The numbers in italics refer to pages with illustrations of the subject.